Lecture Notes in Statistics

Vol. 1: R. A. Fisher: An Appreciation. Edited by S. E. Fienberg and D. V. Hinkley. XI, 208 pages, 1980.

Vol. 2: Mathematical Statistics and Probability Theory. Proceedings 1978. Edited by W. Klonecki, A. Kozek, and J. Rosiński. XXIV, 373 pages, 1980.

Vol. 3: B. D. Spencer, Benefit-Cost Analysis of Data Used to Allocate Funds. VIII, 296 pages, 1980.

Vol. 4: E. A. van Doorn, Stochastic Monotonicity and Queueing Applications of Birth-Death Processes. VI, 118 pages, 1981.

Vol. 5: T. Rolski, Stationary Random Processes Associated with Point Processes. VI, 139 pages, 1981.

Vol. 6: S. S. Gupta and D.-Y. Huang, Multiple Statistical Decision Theory: Recent Developments. VIII, 104 pages, 1981.

Vol. 7: M. Akahira and K. Takeuchi, Asymptotic Efficiency of Statistical Estimators. VIII, 242 pages, 1981.

Vol. 8: The First Pannonian Symposium on Mathematical Statistics. Edited by P. Révész, L. Schmetterer, and V. M. Zolotarev. VI, 308 pages, 1981.

Vol. 9: B. Jørgensen, Statistical Properties of the Generalized Inverse Gaussian Distribution. VI, 188 pages, 1981.

Vol. 10: A. A. McIntosh, Fitting Linear Models: An Application on Conjugate Gradient Algorithms. VI, 200 pages, 1982.

Vol. 11: D. F. Nicholls and B. G. Quinn, Random Coefficient Autoregressive Models: An Introduction. V, 154 pages, 1982.

Vol. 12: M. Jacobsen, Statistical Analysis of Counting Processes. VII, 226 pages, 1982.

Vol. 13: J. Pfanzagl (with the assistance of W. Wefelmeyer), Contributions to a General Asymptotic Statistical Theory. VII, 315 pages, 1982.

Vol. 14: GLIM 82: Proceedings of the International Conference on Generalised Linear Models. Edited by R. Gilchrist. V, 188 pages, 1982.

Vol. 15: K. R. W. Brewer and M. Hanif, Sampling with Unequal Probabilities. IX, 164 pages, 1983.

Vol. 16: Specifying Statistical Models: From Parametric to Non-Parametric, Using Bayesian or Non-Bayesian Approaches. Edited by J. P. Florens, M. Mouchart, J. P. Raoult, L. Simar, and A. F. M. Smith. XI, 204 pages, 1983.

Vol. 17: I. V. Basawa and D. J. Scott, Asymptotic Optimal Inference for Non-Ergodic Models. IX, 170 pages, 1983.

Vol. 18: W. Britton, Conjugate Duality and the Exponential Fourier Spectrum. V, 226 pages, 1983.

Vol. 19: L. Fernholz, von Mises Calculus For Statistical Functionals. VIII, 124 pages, 1983.

Vol. 20: Mathematical Learning Models – Theory and Algorithms: Proceedings of a Conference. Edited by U. Herkenrath, D. Kalin, W. Vogel. XIV, 226 pages, 1983.

Vol. 21: H. Tong, Threshold Models in Non-linear Time Series Analysis. X, 323 pages, 1983.

Vol. 22: S. Johansen, Functional Relations, Random Coefficients and Nonlinear Regression with Application to Kinetic Data. VIII, 126 pages. 1984.

Vol. 23: D. G. Saphire, Estimation of Victimization Prevalence Using Data from the National Crime Survey. V, 165 pages. 1984.

Vol. 24: T. S. Rao, M. M. Gabr, An Introduction to Bispectral Analysis and Bilinear Time Series Models. VIII, 280 pages, 1984.

Vol. 25: Time Series Analysis of Irregularly Observed Data. Proceedings, 1983. Edited by E. Parzen. VII, 363 pages, 1984.

ctd. on inside back cover

Lecture Notes in Statistics

Edited by D. Brillinger, S. Fienberg, J. Gani, J. Hartigan, and K. Krickeberg

43

Barry C. Arnold

Majorization and the Lorenz Order: A Brief Introduction

Springer-Verlag
Berlin Heidelberg New York London Paris Tokyo

Author

Barry C. Arnold
University of California, Department of Statistics
Riverside, CA 92521, USA

AMS Subject Classification (1980): 60E15, 62H15

ISBN 978-0-387-96592-5 ISBN 978-1-4615-7379-1 (eBook)
DOI 10.1007/978-1-4615-7379-1

2147/3140-543210

Preface

My interest in majorization was first spurred by Ingram Olkin's proclivity for finding Schur convex functions lurking in the problem section of every issue of the American Mathematical Monthly. Later my interest in income inequality led me again to try and "really" understand Hardy, Littlewood and Polya's contributions to the majorization literature. I have found the income distribution context to be quite convenient for discussion of inequality orderings. The present set of notes is designed for a one quarter course introducing majorization and the Lorenz order. The inequality principles of Dalton, especially the transfer or Robin Hood principle, are given appropriate prominence.

Initial versions of these notes were used in graduate statistics classes taught at the Colegio de Postgraduados, Chapingo, Mexico and the University of California, Riverside. I am grateful to students in these classes for their constructive critical commentaries.

My wife Carole made noble efforts to harness my free form writing and punctuation. Occasionally I was unmoved by her requests for clarification. Time will probably prove her right in these instances also. Peggy Franklin did an outstanding job of typing the manuscript, and patiently endured requests for innumerable modifications.

Barry C. Arnold
Riverside, California
July, 1986

CONTENTS

CHAPTER 1

INTRODUCTION

The theory of majorization is perhaps most remarkable for its simplicity. How can such a simple concept be useful in so many diverse fields? The plethora of synonyms or quasi-synonyms for variability (diversity, inequality, spread, etc.) suggest that we are dealing with a basic conception which is multifaceted in manifestation and not susceptible to a brief definition which will command universal acceptance. Yet there is an aspect of inequality which comes close to the elusive universal acceptance. The names associated with this identifiable component of inequality are several. Effectively, several authors happened upon the same concept in different contexts. Any list will probably do injustice to some group of early researchers.

In the early accessible English language literature on the subject the names of Muirhead, Lorenz, Dalton and Hardy, Littlewood and Polya stand out. Majorization or Lorenz ordering is the name we attach to the partial order implicitly or explicitly described by these authors. The arena in which inequality measurement was discussed was broad. Lorenz and Dalton did their work in the context of income inequality. It is convenient to use income inequality as our standard example, but the reader is enjoined to recall that any of many other fields of application might serve as well. In fact, concurrent and possible earlier discovery of relevant concepts may well have occurred in other fields.

Hardy, Littlewood and Polya's (1959) book remains a fertile source of mathematical results relating to inequality. The passage of time and perhaps infelicitous choice of notation make that book even less accessible to the student. Marshall and Olkin (1979) have recently written an excellent treatise on majorization. It is much more accessible, but it is overwhelming in scope. Theorems are proved in all extant versions. The present book is designed to give a brief introduction to the material. Hopefully the reader will be stimulated to pursue some topics further in Marshall and Olkin's book or earlier sources. Citations to the literature will be given to help establish the temporal sequence of the development of key ideas. It would, however, be presumptuous to try to improve upon the detailed bibliography supplied by Marshall and Olkin.

The concepts to be discussed are in a sense purely mathematical. Indeed, Hardy, Littlewood and Polya (1959) treated them just that way. Nevertheless, I reiterate my belief that fleshing the ideas out by setting them in some real world context is helpful for motivation and suggestive of useful extensions and generalizations. So, in our introductory sections we will speak of inequality in the income context. In later chapters we will vacillate. Sometimes results will be presented from a purely mathematical point of view. Other times a (sometimes superficial) economic dressing is added. It is hoped that this will encourage the reader to think of analogous applications in his/her field of interest, and that it will not give the erroneous impression that only in economics is inequality, diversity, variability or what have you, of interest.

Before Lorenz's (1905) important paper, there had been several suggestions regarding how inequality might best be measures. Lorenz felt that all of the summary measures then under consideration constituted, in effect, too much condensation of the data. Each provided a snapshot of some aspect of inequality. But as with the blind men and the elephant, the beast itself remained imperfectly mirrored by the unidimensional views. As a more full bodied view, he proposed a curve which has come down to us as the Lorenz curve. Actually it provides almost no condensation of the data. It determines the distribution up to scale. Pushing our blind man and the elephant theme to (or beyond) its limits, it provides a laser picture of the beast with only the actual scale unknown. There are many functionals of distributions which essentially divide the class of all distributions into equivalence classes under change of scale. What is special about the Lorenz curve is that a natural partial order for inequality is derivable from it. This partial order based on nested Lorenz curves, was suggested by Lorenz in his paper and has proved to be the most widely accepted partial ordering relating to inequality. It is, of course, the main subject of this book. Let us look back at its genesis in the publication of the American Statistical Association of 1905.

Lorenz points out negative features of most of the simple summary measures of inequality with a small accolade for Bowley's measure of dispersion. Bowley's measure is touted as the best numerical measure as yet suggested, but it is clear that Lorenz considers it the best of a rotten bunch. He is more concerned with more informative graphical measures. A popular graphical technique was (and remains) available. Vito Pareto proposed plotting log income against the logarithm of the survival function of the distribution. There is good reason to believe the slope of such a chart (called a Pareto chart) will only provide a good measure of inequality when the actual distribution is of the Pareto form [i.e. $P(X > x) = (x/\sigma)^{-\alpha}$, $x > \sigma$]. In fact, the chart will only be linear for such distributions. However, taking logarithms has a way of masking some of the deviations from the Pareto distribution. In addition, much income data does seem to fit the Pareto model reasonably well. Thus, it was generally posible to compare income distributions in terms of the slopes of their approximately linear Pareto charts.

Lorenz rightly sensed that "logarithmic curves are more or less treacherous." He proposed a graphical technique which did not involve such a transformation. To each number between 0 and 100, thought of as a percentage, say t, he proposed to associate the percentage of the total income which accrued to the poorest t percent of the population. He used percentage of income as the abcissa of the point on the curve. Subsequently, it has become customary to reflect Lorenz's curve about the 45° line, i.e., use percentage of income as the ordinate. Lorenz observed that such curves will be typically bow shaped. He proposed the following "rule of interpretation." "As the bow is bent, so concentration" (i.e., inequality) "increases." Presumably this rule of interpretation is self evident, for Lorenz gives no hint of justification. Was this insight or seredipity? Muirhead's (1903) work was already published, but the economic interpretation provided by Dalton's transfer principles was still 15 years away. It seems doubtful that Lorenz had any precise mathematical formulation for his rule of interpretation. It seemed logical to him, and it has survived well simply because it really does capture important aspects of inequality. But Dalton's work was needed to clarify this.

Hugh Dalton, later to become chancellor of the exchequer in England for the Labor government elected in 1946, was a practical economist. Perhaps because of his practicality, he demanded precision in definition. He did not achieve a precise definition of inequality in his pioneering work of 1920, but he did point out some key ideas regarding desirable properties of inequality measures. He tried to isolate operations on income distributions which would "clearly" increase inequality and then strove to identify measures of inequality which would be monotone under such transformations. One might quibble that we are merely replacing one difficult problem by another possibly more difficult one, i.e., how can we agree on what transformations will clearly increase inequality? It would appear in retrospect that only one of Dalton's inequality principles commands wide, almost universal, acceptance. It is known as the Pigou-Dalton transfer principle, but in this book we will use the more evocative name of the Robin Hood axiom. When Robin and his merry hoods performed an operation in the woods they took from the rich and gave to the poor. The Robin Hood principle asserts that this decreases inequality (subject only to the obvious constraint that you don't take too much from the rich and turn them into poor.) Most would agree that a Robin Hood operation decreases inequality. If we are to judge by names, since a Robin Hood heist is known as a progressive transfer, it appears that most approve of such Robin Hood actions. Presumably the sheriff of Nottingham was not a right thinker in this regard. It turns out that the Robin Hood axiom is intimately related to the order proposed by Lorenz. A chance for one last Sherwood metaphor is provided by Lorenz's comment about inequality increasing as the bow is bent. Unfortunately Robin's bow bending efforts decrease rather than increase inequality, but we can forgive Lorenz his lack of prescience here.

Dalton's second principle was that multiplication of incomes by a constant greater than one should decrease inequality. Most popular inequality measures do not satisfy this criterion. In fact, arguments regarding reexpression of incomes in new currency units suggest that a desirable property of inequality measures would be scale invariance. We will accept this revised second principle. Dalton's third principle, which may be simply phrased as "giving everyone a dollar will decrease inequality," is, as we shall see, a consequence of Principle 1 and revised Principle 2 (scale invariance). His fourth principle is that the measure of inequality should be "invariant under cloning". Specifically, consider a population of n individuals and a related population of kn individuals which consists of k identical copies (with regard to income) of each of the n individuals in the original population. The population of size kn can be called the cloned population. Principle four requires that the measure of inequality yield the same value for the cloned population as it does for the original population. Mathematically this requires that the measure of inequality for the population should be a function of the sample distribution function of the population. Most common measures of inequality satisfy this last principle.

Now how are we to get from Dalton and Lorenz or, if you will, Robin Hood and Lorenz's bow to the titular topic of this book, namely, majorization? And where does Muirhead fit in? It is not easy to guess dates. But one might speculate that the ideas contained in Hardy, Littlewood and Polya's (1929) paper had been formulated several years earlier, perhaps just when Dalton was enunciating his principles. Be that as it may, their brief 8 page paper in the Messenger of Mathematics contains the fetus of majorization theory fully formed and, indeed, more. How Topsy has grown; from part of Hardy, Littlewood and Polya's 8 pages to all of Marshall and Olkin's 520. Perhaps the main reason for the slow recognition of the nexus of the concepts of Lorenz ordering and majorization is to be found in the mathematician's proclivity to arrange numbers in decreasing order as opposed to the statistician's tendency to use increasing order! So that both may feel at home, we will phrase our definition both ways.

What then is majorization? We could continue to speak of incomes in a finite population, but it is a convenient time to drop the economic trappings in favor of clear mathematical statements. Majorization is a partial order defined on the positive orthant of n dimensional Euclidean space, to be denoted by $\mathbb{R}_n^+$. For a vector

$\underline{x} \in \mathbb{R}_n^+$ we denote its (increasing) order statistic by $(x_{1:n}, x_{2:n}, \ldots, x_{n:n})$, i.e., the x_i's written in increasing order. Its decreasing order statistic will be denoted by $x_{(1:n)}, x_{(2:n)}, \ldots, x_{(n:n)}$. Thus $x_{1:n}$ is the smallest of the x's, while $x_{(1:n)}$ is the largest. Of course, $x_{i:n} = x_{(n-i+1:n)}$. Along with most mathematicians, HLP (an acronym for Hardy, Littlewood and Polya that we will henceforth adopt) used the decreasing order statistics in their original definition.

Definition: Let $\underline{x}, \underline{y} \in \mathbb{R}_n^+$. We will say that $\underline{x}$ majorizes $\underline{y}$ and write $\underline{x} \geq_M \underline{y}$ if

$$\sum_{i=1}^{k} x_{(i:n)} \geq \sum_{i=1}^{k} y_{(i:n)}, \quad k = 1,2,\ldots,n-1 \tag{1.1}$$

and $\sum_{i=1}^{n} x_{(i:n)} = \sum_{i=1}^{n} y_{(i:n)}$ [equivalently, if $\sum_{i=1}^{k} x_{i:n} \leq \sum_{i=1}^{k} y_{i:n}$, $k = 1,2,\ldots,n-1$] and $\sum_{i=1}^{n} x_i = \sum_{i=1}^{n} y_i$.

Now how can we write a book about that? First of all, let us see what it says in the income context. Suppose we have two populations of n individuals each, with equal total incomes assumed without loss of generality to be 1. If we plot the points $(k/n, \sum_{i=1}^{k} x_{i:n})$ and $(k/n, \sum_{i=1}^{k} y_{i:n})$, we see that $\underline{x} \geq_M \underline{y}$ if and only if the Lorenz curve of $\underline{y}$ is wholly nested within that of $\underline{x}$. Thus, $x \geq_M \underline{y}$ if and only if $\underline{x}$ is more unequal than $\underline{y}$ in the ordering proposed by Lorenz (by bow bending). In $\mathbb{R}_n^+$ we may thus speak of majorization and the Lorenz ordering essentially interchangeably.

Where does Robin Hood fit into this majorization business? When Robin Hood does his work, he performs a progressive transfer. The new income distribution $\underline{y}$ is related to the old distribution $\underline{x}$ by

$$\underline{y} = P\underline{x}$$

where P is a very simple doubly stochastic matrix. Using Muirhead's (1903) paper HLP (1959) show that $\underline{x} \geq_M \underline{y}$ (or $\underline{x} \geq_L \underline{y}$, if we wish to speak of the Lorenz order), if and only if $\underline{y}$ can be obtained from $\underline{x}$ by Robin Hood in a finite number of operations. So if we want an inequality measure to honor Dalton's first principle, then it must be a function on $\mathbb{R}_n^+$ which preserves the majorization partial order. Such functions are known as Schur convex functions.

Schur convex functions antedate the concept of majorization by a decade. Schur (1923) spoke of an averaging of $\underline{x}$ to be any $\underline{y}$ which was of the form $\underline{y} = P\underline{x}$ where P is doubly stochastic. He identified the class of functions which preserved the partial ordering defined by such averaging. HLP (1959) showed that Schur's averaging partial order was equivalent to majorization. Actually Muirhead (1903) had considered what was to become known as majorization in a much earlier paper, although he considered it as a partial order on vectors of non-negative integers.

As HLP (1959) noted in their book, there is a good reason to expect that any theorem which provided inequalities for vectors in $\mathbb{R}_n^+$ using summation may well have a generalization for non-negative functions involving integration. The analog of the order statistic for a vector in $\mathbb{R}_n^+$ is provided by the increasing rearrangement of a non-negative function. They did not provide the necessary extension of Schur's averaging, but even that is possible using the concept of a balayage. One can, in fact, parlay the exercise into even more abstract settings, but we will satisfy ourselves here with the mere mention of such possibilities.

One important extension which does not take us into esoteric territory, but nevertheless into a thicket of problems, involves multivariate majorization. The basic ambiguity is a consequence of the fact that there really is no compelling ordering of vectors to play the role that the order statistics played in the development of (univariate) majorization. Inevitably, stochastic versions of majorization had to evolve. Some discussion of these concepts will be provided in the latter part of Chapter 5. Our penultimate chapter explores, albeit superfically, the relationship between majorization and several related partial orderings including stochastic dominance. The concluding chapter catalogs a variety of applications of majorization.

As an entertainment, one might attempt to enumerate how many surrogates of the majorization partial ordering are described in this book. Important ones are introduced on pages 14 and 33. But, there are others sprinkled around.

CHAPTER 2

MAJORIZATION IN $\mathbb{R}_n$

As mentioned in Chapter 1, the name majorization appears first in HLP (1959). The idea had appeared earlier (HLP, 1929) although unchristened. Muirhead who dealt with $\mathbb{Z}_n^+$ (i.e. vectors of non-negative integers) already had identified the partial order defined in (1.1) (i.e. majorization). But he, when he needed to refer to it, merely called it "ordering." Perhaps it took the insight of HLP to recognize that little of Muirhead's work need necessarily be restricted to integers, but the key ideas including Dalton's transfer principle were already present in Muirhead's paper. If there was anything lacking in Muirhead's development, it was motivation for the novel results he obtained. He did exhibit the arithmetic-geometric mean inequality as an example of his general results, but proofs of that inequality are legion. If that was the only use of his "inequalities of symmetric algebraic functions of n letters", then they might well remain buried in the Edinburgh proceedings. HLP effectively rescued Muirhead's work from such potential obscurity. In the present book theorems will be stated in generality comparable to that achieved by HLP and will be ascribed to those authors. Muirhead's priority will not be repeatedly asserted. HLP restricted attention to $\mathbb{R}_n^+$, but the restriction to the positive orthant can and will be often dispensed with. First let us establish the relationship between majorization as defined by HLP (i.e., (1.1)) and averaging as defined by Schur. Recall that $\underline{x}$ is an average of $\underline{y}$ in the Schur sense if $\underline{x} = P\underline{y}$ for some doubly stochastic matrix P.

There are certain very simple linear transformations, $\underline{x} = A\underline{y}$, for which it is obvious that $\underline{x} \leq_M \underline{y}$. The very simplest case is when

A is a permutation matrix. But, there are other quite straightforward cases. Going back to our Robin Hood scenario of Chapter 1, we would like to show that the linear transform associated with a Robin Hood operation leads to a new vector $\underline{x}$ which is majorized by the old vector $\underline{y}$. In order to obtain results valid in $\mathbb{R}_n$ rather than just $\mathbb{R}_n^+$, it is convenient to admit the possibility of negative incomes in our financial scenario. To begin with we consider a special kind of Robin Hood operation in which money is taken from a relatively rich individual i and given to the individual whose wealth is immediately below that of individual i in the ranking of wealth. Call this an elementary Robin Hood operation.

Because of the fact that permutation preserves majorization, we may, without loss of generality, assume that our vector $\underline{x}$ has coordinates arranged in inreasing order. Now, what we want to show is that every Robin Hood operation is equivalent to a series of elementary Robin Hood operations and permutations and, then that every multiplication by a doubly stochastic matrix is equivalent to a finite series of Robin Hood operations (and thus to a longer series of elementary Robin Hood operations and permutations). Then if the elementary Robin Hood operations induce majorization, so does multiplication by a doubly stochatic matrix.

First let us ascertain that an elementary Robin Hood operation induces majorization. The old and new incomes (possibly negative) are related by

$$\underline{x} = A\underline{y} \tag{2.1}$$

where, for some i and some $\lambda \varepsilon [0 , \frac{1}{2}]$,

$$a_{i,i} = 1 - \lambda, \qquad a_{i,i+1} = \lambda,$$

$$a_{i+1,i} = \lambda, \qquad a_{i+1,i+1} = 1 - \lambda \tag{2.2}$$

and

$a_{jk} = \delta_{jk}$ otherwise,

where δ_{jk} denotes the usual Kronecker delta. Note that we require $\lambda \in [0, \frac{1}{2}]$ in order to have Robin Hood's operation not disturb the ordering of the vector. It is evident from the basic definition (1.1) that with A defined by (2.2), and $\underline{y}$ and $\underline{x}$ related by (2.1) we have $\underline{x} \leq_M \underline{y}$. Now a more general Robin Hood operation involves taking the income of a relatively poor individual and a relatively rich individual, perhaps individuals i and j in the increasing ranking, and redividing their combined wealth among them, subject only to the constraint that individual i not become more rich than individual j. The effect of such an operation is to raise up the Lorenz curve (the plot of points $(k/n, \sum_{i=1}^{k} x_{i:n})$). The effect of an elementary Robin Hood operation is to raise the Lorenz curve at one point. It is a simple matter to verify that one may move from a given Lorenz curve to a higher one by successively raising the curve at individual points subject to the constraint that at no time is the convexity of the Lorenz curve destroyed. A simple example will exhibit the manner in which this may be achieved.

Consider a population of 8 individuals whose ordered wealths are

2, 3, 5, 11, 13, 18, 23, 25 . (A)

Suppose that Robin Hood takes 8 from the individual with wealth 23 and gives it to the individual with wealth 5. The resulting set of ordered wealths is

2, 3, 11, 13, 13, 15, 18, 25 . (B)

It is not difficult to reach (B) from (A) using only elementary Robin Hood operations. For example one could successively transform the set of ordered wealths as follows.

2,	3,	5,	11,	13,	18,	23,	25
2,	3,	5,	11,	13,	20.5,	20.5,	25
2,	3,	5,	11,	15.5,	18,	20.5,	25
2,	3,	5,	13,	13.5,	18,	20.5,	25
2,	3,	9,	9,	13.5,	18,	20.5,	25
2,	3,	9,	9,	15.5,	16,	20.5,	25
2,	3,	9,	12,	12.5,	16,	20.5,	25
2,	3,	9,	12,	13,	15.5,	20.5,	25
2,	3,	9,	12,	13,	18,	18,	25
2,	3,	10,	11,	13,	18,	18,	25
2,	3,	10,	12,	12,	18,	18,	25
2,	3,	10,	12,	15,	15,	18,	25
2,	3,	11,	11,	15,	15,	18,	25
2,	3,	11,	13,	13,	15,	18,	25

Each line above was obtained from the preceding line by an elementary Robin Hood operation. The reader might ponder on the problem of determining the minimal number of elementary Robin Hood operations required to duplicate a given general Robin Hood operation. The sequence described above is not claimed to be parsimonious with regard to the number of steps used. Duplication of a general Robin Hood operation may require a countable number of elementary Robin Hood operations (see exercise 1).

But now we may make a remarkable observation. A vector $\underline{x}$ majorizes $\underline{y}$, by definition, if the Lorenz curve of $\underline{y}$ is obtained from that of $\underline{x}$ by raising it at one or several points. It is evident that this too can be accomplished one point at a time (subject to the restriction that convexity is not violated). Let $\mathcal{E}$ be the class of all n×n doubly stochastic matrices corresponding to elementary Robin Hood operations, and let $\mathcal{P}$ be the class of all n×n permutation matrices. It is then clear that if $\underline{x} \geq_M \underline{y}$, then $\underline{y}$ can be obtained from $\underline{x}$ by a string of elementary Robin Hood operations and/or relabelings. Thus $\underline{y} = Q\underline{x}$ where Q is a product of a countable number of matrices chosen from the class of doubly stochastic matrices $\mathcal{E} \cup \mathcal{P}$. Q is then necessarily itself doubly stochastic. Thus we have proved that if $\underline{x} \geq_M \underline{y}$, then $\underline{y} = P\underline{x}$ for some doubly stochastic matrix P. The converse statement was proved in HLP essentially as follows.

Suppose that $\underline{y} = P\underline{x}$ for some doubly stochastic matrix P where without loss of generality $\underline{x}$ and $\underline{y}$ are in ascending order. Then for any $m \leq n$, we may define

$$k_j = \sum_{i=1}^{m} p_{ij} \leq 1; \qquad j = 1,2,\ldots,n,$$

and we have

$$\begin{aligned}
\sum_{i=1}^{m} y_i &= \sum_{j=1}^{n} k_j x_j \\
&\geq \sum_{j=1}^{m-1} k_j x_j + x_m \left(m - \sum_{j=1}^{m-1} k_j\right) \\
&\quad \left(\text{since } \sum_{j=1}^{n} k_j = m\right) \\
&\geq \sum_{j=1}^{m-1} k_j x_j + x_m + \sum_{j=1}^{m-1} (1-k_j) x_j \\
&\quad (\text{since } k_j \leq 1) \\
&= \sum_{j=1}^{m} x_j .
\end{aligned}$$

Thus $\underline{x} \geq_M \underline{y}$. We may consequently state

<u>Theorem 2.1</u> (HLP). $\underline{x} \geq_M \underline{y}$ if and only if $\underline{y} = P\underline{x}$ for some doubly stochastic matrix P.

The matrix P referred to in Theorem 2.1 is not necessarily unique. For a given $\underline{x}$, $\underline{y}$ the class of matrices P for which $y = P\underline{x}$ is convex, but it can be a singleton set or in other cases may be nontrivial and have several extreme points.

An alternative characterization of majorization is then available if we utilize Birkhoff's (1946) observation that the class of n×n doubly stochastic matrices coincides with the convex hull of the set of n×n permutation matrices. We thus may state that $\underline{x} \geq_M \underline{y}$ if and only if $\underline{y} = \sum_{\ell=1}^{k} \gamma_\ell P_\ell \underline{x}$ for some set of permutation matrices $P_1, P_2, \ldots, P_k$ and some set of γ_ℓ's satisfying $\gamma_\ell \geq 0$, $\ell = 1,2,\ldots,k$ and $\sum_{\ell=1}^{k} \gamma_\ell = 1$. Farahat and Mirsky (1960) showed that k can be chosen to be $n^2 - 2n + 2$ and that, in general, no smaller number will suffice. An alternative statement of this fact is that for a given vector $\underline{x}$ the class of vectors $\underline{y}$ majorized by $\underline{x}$ is the convex hull of the set of points obtained by rearranging the coordinates of $\underline{x}$.

Functions which preserve the majorization partial order form a large class. They were first studied by Schur (1923) and are, in his honor, called Schur convex functions. We may join with Marshall and Olkin in lamenting the use of the term convex rather than monotone, increasing, order preserving or, more modishly, isotonic, but we will persist in using the name Schur convex. Many famous inequalities are readily proved by focussing on a particular Schur convex function. If one thumbs through the problem sections of the American Mathematical Monthly one is struck by the frequency with which inequalities are proved as a consequence of the Schur convexity of some judiciously chosen function. For example, Chapter 8 of Marshall and Olkin (1979) catalogs a plethora of such results in a geometric setting (many dealing with features of triangles). Not every analytic inequality is a consequence of the Schur convexity of some function, but enough are to make familiarity with majorization/Schur-convexity a necesary part of the required background of a respectable mathematical analyst. After that build-up, we had better get quickly to the matter of defining the required concepts.

<u>Definition. 2.2.</u> Let $A \subset \mathbb{R}_n$. A function $g:A \to \mathbb{R}$ is said to be <u>Schur convex</u> on A if $g(\underline{x}) \le g(\underline{y})$ for every pair $\underline{x},\underline{y} \in A$ for which $\underline{x} \le_M \underline{y}$.

The function g alluded to in definition 2.2 is said to be strictly Schur convex if it is Schur convex and if $\underline{x} \le_M \underline{y}$, and $\underline{x}$ not a permutation of $\underline{y}$ together imply $g(\underline{x}) < g(\underline{y})$.

As an example consider the function

$$g(\underline{x}) = \sum_{i=1}^{n} x_i^2 \tag{2.3}$$

defined on $\mathbb{R}_n$. We claim that this function is Schur convex. In verifying this fact we will use a couple of tricks of the trade. The task of determining Schur-convexity is simplified immediately by the observation that a Schur convex function must be a symmetric function of its arguments. More specifically, if g is Schur convex on A and if $\underline{x}$ and $\Pi\underline{x}$ are in A for some permutation matrix Π, then $g(\underline{x}) = g(\Pi\underline{x})$ (since, clearly, $\underline{x} \ge_M \Pi\underline{x}$ and $\Pi\underline{x} \ge_M \underline{x}$). Another simplification hinges upon the fact that if $\underline{x} \ge_M \underline{y}$, then $\underline{y}$ can be obtained from $\underline{x}$ by a finite number of Robin Hood operations. Such operations only involve two coordinates at a time. So we only need to check whether $g(\underline{x}) \le g(\underline{y})$ whenever $\underline{x} \le_M \underline{y}$ and $\underline{x}$ and $\underline{y}$ differ only in two coordinates which by symmetry may without loss of generality be taken to be the first two.

Thus, in order to verify Schur convexity of the function (2.3), we need to verify that if

$$(\alpha_1,\alpha_2,\alpha_3,\ldots,\alpha_n)' \le_M (\beta_1,\beta_2,\alpha_3,\ldots,\alpha_n)' \tag{2.4}$$

(where necessarily $\alpha_1 + \alpha_2 = \beta_1 + \beta_2$ and without loss of generality $\alpha_1 \le \alpha_2$ and $\beta_1 \le \beta_2$), then $\alpha_1^2 + \alpha_2^2 + \ldots + \alpha_n^2 \le \beta_1^2 + \beta_2^2 + \alpha_3^2 + \ldots + \alpha_n^2$, i.e.,

$$\alpha_1^2 + \alpha_2^2 \le \beta_1^2 + \beta_2^2 .$$

Without loss of generality $\alpha_1 + \alpha_2 = \beta_1 + \beta_2 = 1$, and we may write α for α_1 and β for β_1. We then wish to verify that if

$$\frac{1}{2} \geq \alpha \geq \beta$$

(as implied by (2.4)), then

$$\alpha^2 + (1-\alpha)^2 \leq \beta^2 + (1-\beta)^2 .$$

This is, however, obvious since the function $z^2 + (1-z)^2$ has a non-positive derivative on $[-\infty, \frac{1}{2}]$.

Let us denote by 0_n the subset of $\mathbb{R}_n$ in which the coordinates are in increasing order. Thus

$$0_n = \{\underline{x}: x_1 \leq x_2 \leq \cdots \leq x_n\} . \tag{2.5}$$

If we wish to check for Schur convexity of a function g defined on a symmetric set A, we need only check for symmetry on A and Schur convexity on $0_n \cap A$. Since, in many applications the domains of putative Schur convex functions are indeed symmetric, it is often sufficient to restrict attention to 0_n or subsets thereof.

To characterize Schur convexity on subsets of 0_n, we return to the Robin Hood scenario. Thinking of the coordinates of 0_n as the ordered wealths of n individuals in Sherwood Forest (allowing negative wealth so as not to restrict attention to $\mathbb{R}_n^+$), we may recall that (in 0_n) $\underline{x} \leq_M \underline{y}$ if and only if $\underline{x}$ can be obtained from $\underline{y}$ by a countable number of elementary Robin Hood operations. Such operations involve a transfer of funds from one individual to the next richest individual without disturbing the ordering of individuals. To be Schur convex on 0_n then, a function g need only be appropriately monotone with respect to such elementary Robin Hood operations (cf. Lemma A.2 in Marshall and Olkin (1979, p. 55)). Monotonicity is particularly easy to verify in differentiable cases by inspection of the sign of the appropriate partial derivative. Schur's (1923) famous sufficient condition for Schur convexity involves such partial derivatives. Although Schur's condition has proved remarkably useful, it cannot be used to deal with less regular Schur convex functions. The existence of unfriendly Schur convex functions is alluded to in exercise 3.

Summarizing the results described above we have

<u>Theorem 2.3</u>: Suppose $g: \mathcal{O}_n \to \mathbb{R}$. g is Schur convex if and only if for every $\underline{x} \in \mathcal{O}_n$ and every $i = 1,2,\ldots,n-1$, the function

$$g(\underline{x} + \varepsilon \underline{v}_i)$$

is a non-increasing function of ε for all ε such that $\underline{x} + \varepsilon \underline{v}_i \in \mathcal{O}_n$. The vector $\underline{v}_i$ is a vector with a 1 in the ith coordinate and a -1 in the (i+1)st coordinate, all other coordinates being 0.

[If the functions $g(\underline{x} + \varepsilon \underline{v}_i)$ are strictly decreasing, then we characterize strict Schur convexity.]

<u>Theorem 2.4</u>: Suppose $g: \mathcal{O}_n \to \mathbb{R}$ is continuous with continuous partial derivatives (on the interior of $\mathcal{O}_n$). g is Schur convex on $\mathcal{O}_n$ if and only if for every $\underline{x}$ in the interior of $\mathcal{O}_n$ and every $j = 1,2,\ldots,n-1$,

$$\frac{\partial}{\partial x_j} g(\underline{x}) \le \frac{\partial}{\partial x_{j+1}} g(\underline{x})$$

[i.e., if $\nabla g(\underline{x}) \in \mathcal{O}_n \ \forall\, \underline{x} \in \text{int}(\mathcal{O}_n)$].

<u>Proof</u>: Follows by legitimately differentiating the $g(\underline{x} + \varepsilon \underline{v}_i)$'s.

<u>Theorem 2.5</u>: [Schur's condition.] Suppose $g: I^n \to \mathbb{R}$ is continuously differentiable where I is an open interval. g is Schur convex if and only if g is symmetric and for every $i \neq j$

$$(x_i - x_j)\left[\frac{\partial}{\partial x_i} g(\underline{x}) - \frac{\partial}{\partial x_j} g(\underline{x})\right] \ge 0 \ \forall\, \underline{x} \in I^n . \tag{2.6}$$

<u>Proof</u>: The result follows from Theorem 2.4 and the assumed symmetry of $g(\underline{x})$. In fact, using the assumed symmetry of $g(\underline{x})$, one only need verify (2.6) for the particular case $(i,j) = (1,2)$.

If we return to our example $g(\underline{x}) = \sum_{i=1}^{n} x_i^2$ defined on $\mathbb{R}_n$, symmetry is evident and we see that Schur's condition (2.6) takes the form

$$(x_1 - x_2)(2x_1 - 2x_2) \geq 0 .$$

Actually this particular Schur convex function provides an example of an important class of Schur convex functions, the separable convex class.

<u>Definition 2.6</u>. A function $g: I^n \to \mathbb{R}$ (where I is an interval) is said to be separable convex if g is of the form

$$g(\underline{x}) = \sum_{i=1}^{n} h(x_i) \tag{2.7}$$

where h is a convex function on I. Recall that h is convex on I if $h(\alpha x + (1-\alpha)y) \leq \alpha h(x) + (1-\alpha)h(y)$ for every $x,y \in I$ and every $\alpha \in [0,1]$.

Schur (1923) and HLP observed that any separable convex function is Schur convex. If h is differentiable (as "most" convex functions are), then Schur's condition (2.6) is readily verified. More generally it can be verified by considering the effect of an elementary Robin Hood operation which involves an averaging of two coordinates (this is exercise 4).

An interesting example of a separable convex function is provided by the function

$$g(\underline{x}) = -\sum_{i=1}^{n} \log x_i \tag{2.8}$$

defined on $\mathbb{R}_n^+$. The function $-\log x$ is readily shown to be convex. A consequence of the Schur convexity of (2.8) is the celebrated arithmetic-geometric mean inequality. For any $\underline{x} \in \mathbb{R}_n^+$ it is evident that

$$(x_1,x_2,\ldots,x_n)' \geq_M (\bar{x},\ldots,\bar{x})' \tag{2.9}$$

where $\bar{x} = \frac{1}{n} \sum_{i=1}^{n} x_i$ is the arithmetic mean of the x_i's (Robin Hood can clearly eventually make everyone's wealth equal). If we evaluate the Schur convex function defined in (2.8) at each of the points in $\mathbb{R}_n^+$ referred to in (2.9), we eventually conclude that

$$\left(\prod_{i=1}^{n} x_i\right)^{\frac{1}{n}} \le \left(\frac{1}{n} \sum_{i=1}^{n} x_i\right) , \tag{2.10}$$

i.e., the geometric mean cannot exceed the arithmetic mean. Myriad proofs of (2.10) exist, a remarkable number of which can be presented in a majorization framework. As remarked at the beginning of the chapter, Muirhead (1903) provided an early example of such a proof.

Not every Schur convex function is separable convex. This is not a transparent result. A famous example of a non-separable Schur convex function is the Gini index. This is defined on $\mathbb{R}_n^+$ as follows

$$g(\underline{x}) = \left[\sum_{i=1}^{n} (2i-n-1)x_{i:n}\right] / \left(\sum_{i=1}^{n} x_i\right) . \tag{2.11}$$

Another such example is provided by

$$g(\underline{x}) = \left(\sum_{i=1}^{n} x_i\right)\left(\sum_{i=1}^{n} x_i^2\right) . \tag{2.12}$$

The elementary symmetric functions

$$\sum_{i=1}^{n} x_i, \quad \sum_{i \ne j} x_i x_j, \quad \sum_{i \ne j \ne k} x_i x_j x_k \text{ etc.} \tag{2.13}$$

are examples of Schur concave functions (g is Schur concave if $-g$ is Schur convex). Except for the first one, $\left(\sum_{i=1}^{n} x_i\right)$, they are clearly not separable concave.

The class of Schur convex functions is closed under a variety of operations. For example consider m Schur convex functions $g_1,\ldots,g_m$ each defined on some set $A \subset \mathbb{R}_n$. It is easily established that the following functions are necessarily Schur convex on A.

$$h_1 = \sum_{i=1}^{m} c_i g_i \quad \text{where} \quad c_i \geq 0 ,$$

$$h_2 = \prod_{i=1}^{m} c_i g_i \quad \text{where} \quad c_i \geq 0 \quad \text{and} \quad g_i \geq 0 ,$$

$$h_3 = \min_i c_i g_i \quad \text{where} \quad c_i \geq 0 , \tag{2.14}$$

$$h_4 = \max_i c_i g_i \quad \text{where} \quad c_i \geq 0 .$$

It is also evident that the class of Schur convex functions is closed under pointwise convergence. In all the above cases Schur convexity is verifiable directly from Definition 2.2.

Another useful way to construct Schur convex functions (or way to recognize Schur convexity) involves marginal transformations before applying a known Schur convex function. Specifically suppose that g is a Schur convex function defined on I^n (I an interval in $\mathbb{R}$) and that h is a convex function defined on $\mathbb{R}$. Define

$$\phi(\underline{x}) = g(h(x_1),\ldots,h(x_n)) . \tag{2.15}$$

A sufficient condition that ϕ be Schur convex is that g be non-decreasing in addition to being Schur convex. We could use this observation to verify yet again the Schur convexity of the function $\sum_{i=1}^{n} x_i^2$ (since evidently $h(x) = x^2$ is convex and $\sum_{i=1}^{n} x_i$ is Schur convex and increasing).

In statistical applications several important Schur convex functions are constructed via integral transforms. We will satisfy ourselves with a representative example (due to Marshall and Olkin (1974)).

Lemma 2.7. Let g be an integrable Schur convex function on $\mathbb{R}_n$, and suppose that $A \subset \mathbb{R}_n$ satisfies:

$$\text{If } \underline{x} \in A \text{ and } \underline{y} \leq_M \underline{x} \text{ then } \underline{y} \in A . \tag{2.16}$$

It follows that

$$\phi(\underline{x}) = \int_{A+\underline{x}} g(\underline{y})d\underline{y}$$

is a Schur convex function of $\underline{x}$ on $\mathbb{R}_n$.

Since measurable Schur concave functions are approximable by simple Schur concave functions (linear combinations of indicator functions), the above Lemma can be parlayed (as Marshall and Olkin did) into a proof of the following.

Theorem 2.8. If g is Schur convex on $\mathbb{R}_n$ and f is Schur concave on $\mathbb{R}_n$, then

$$\phi(\underline{x}) = \int_{\mathbb{R}_n} f(\underline{x} - \underline{y})\, g(\underline{y})\, d\underline{y} \tag{2.17}$$

is Schur convex on $\mathbb{R}_n$ provided that the integral in (2.17) exists for all $\underline{x} \in \mathbb{R}_n$.

The key observation in proving Theorem 2.8 by way of Lemma 2.7 is that (2.16) is equivalent to the statement that the indicator function of A is Schur concave.

The following direct proof of Theorem 2.8 is attributed by Marshall and Olkin to Proschan and Cheng. Without loss of generality, assume n=2 (since we only need to consider elementary Robin Hood operations). Additionally we need only consider small heists. So for ε small, consider $(x_1 + \varepsilon, x_2 - \varepsilon) \leq_M (x_1,x_2)$ where $x_1 < x_2$. We have

$$\begin{aligned}
&\phi(x_1,x_2) - \phi(x_1 + \varepsilon, x_2 - \varepsilon) \\
&\quad = \int_{\mathbb{R}_2} f(x_1 - y_1, x_2 - y_2)\, g(y_1,y_2)\, dy_1dy_2 \\
&\qquad - \int_{\mathbb{R}_2} f(x_1 + \varepsilon - y_1, x_2 - \varepsilon - y_2)\, g(y_1,y_2)\, dy_1dy_2 \\
&\quad = \int_{\mathbb{R}_2} f(u_1,u_2 + \varepsilon)\, g(x_1 - u_1, x_2 - \varepsilon - u_2)\, du_1du_2 \\
&\qquad - \int_{\mathbb{R}_2} f(u_1 + \varepsilon, u_2)\, g(x_1 - u_1, x_2 - \varepsilon - u_2)\, du_1du_2 .
\end{aligned}$$

Since f is symmetric we can write

$$\int_{u_1 \le u_2}[f(u_1,u_2+\varepsilon) - f(u_1+\varepsilon,u_2)]g(x_1-u_1,x_2-\varepsilon-u_2)du_1du_2$$

$$= \int_{u_1 \le u_2}[f(u_2+\varepsilon,u_1) - f(u_2,u_1+\varepsilon)]g(x_1-u_1,x_2-\varepsilon-u_2)du_1du_2$$

$$= \int_{u_1 \ge u_2}[f(u_1+\varepsilon,u_2) - f(u_1,u_2+\varepsilon)]g(x_1-u_2,x_2-\varepsilon-u_1)du_1du_2$$

(relabelling u_1 and u_2). Consequently,

$$\phi(x_1,x_2) - \phi(x_1+\varepsilon,x_2-\varepsilon)$$
$$= \int_{u_1 \ge u_2}[f(u_1,u_2+\varepsilon) - f(u_1+\varepsilon,u_2)][g(x_1-u_1,x_2-\varepsilon-u_2)$$
$$- g(x_1-u_2,x_2-\varepsilon-u_1)]du_1du_2 .$$

However for $u_1 \ge u_2$ we have

$$(u_1,u_2+\varepsilon) \le_M (u_1+\varepsilon,u_2)$$

and

$$(x_1-u_2,x_2-\varepsilon-u_1) \le_M (x_1-u_1,x_2-\varepsilon-u_2) .$$

Consequently, by the Schur concavity of f and the Schur convexity of g, the integrand is non-negative and the Schur convexity of ϕ is verified.

Other useful examples of majorization involving integral transformations are to be found in the work of Hollander, Proschan and Sethuraman (1977) in the context of functions "decreasing in transposition".

The definition of majorization involves ordering the components of vectors in $\mathbb{R}_n$. If n is not large, this is not an arduous task. If n is large, then it becomes desirable to determine sufficient conditions for majorization which do not involve ordering. Of course a sufficient condition that $\underline{x} \le_M \underline{y}$ is that $g(\underline{x}) \le g(\underline{y})$ for every Schur convex function g. This is just the contrapositive of the definition of Schur convexity. This does not involve ordering, but it does involve checking a vast number of functions. The cardinality of the set of Schur convex functions may be larger than your first guess in light of the question at the end of exercise 3. Surely we do not have to check whether $g(\underline{x}) \leq g(\underline{y})$ for every Schur convex function g. The pathological ones alluded to in exercise 3 certainly don't have to be checked (why?). In 1929 HLP verified that one need only check separable convex functions in order to verify majorization. In fact it suffices to check only a particularly simple subclass of the separable convex functions. In what follows we use the notation $a^+ = \max(0,a)$.

Theorem 2.9 (HLP, Karamata): $\underline{x} \leq_M \underline{y}$ if and only if $\sum_{i=1}^n h(x_i) \leq \sum_{i=1}^n h(y_i)$ for every (continuous) convex function $h: \mathbb{R} \to \mathbb{R}$.

Theorem 2.10 (HLP): $\underline{x} \leq_M \underline{y}$ if and only if $\sum_{i=1}^n x_i = \sum_{i=1}^n y_i$ and $\sum_{i=1}^n (x_i-c)^+ \leq \sum_{i=1}^n (y_i-c)^+$ for every $c \in \mathbb{R}$.

It is evidently sufficient to prove Theorem 2.10, since $h(x) = x$ and $h(x) = (x-c)^+$ are continuous convex functions. It is also then evident that Theorem 9 remains valid with or without the parenthetical word "continuous".

Proof of 2.10: One implication is trivial since the functions $\sum_{i=1}^n x_i$ and $\sum_{i=1}^n (x_i-c)^+$ are separable convex and hence Schur convex. The proof of the converse is also straightforward. Merely set c successively equal to $y_{(1:n)}, y_{(2:n)}, \ldots, y_{(n:n)}$. Thus, in terms of decreasing order statistics,

$$\left(\sum_{i=1}^k x_{(i:n)}\right) - k y_{(k:n)} \leq \sum_{i=1}^k \left(x_{(i:n)} - y_{(k:n)}\right)^+$$

$$\leq \sum_{i=1}^n \left(x_i - y_{(k:n)}\right)^+$$

$$\leq \sum_{i=1}^n \left(y_i - y_{(k:n)}\right)^+ \text{ by hypothesis}$$

$$= \sum_{i=1}^{k} \left(y_{(i:n)} - y_{(k:n)}\right)$$

$$= \sum_{i=1}^{k} y_{(i:n)} - ky_{(k:n)} .$$

Adding $ky_{(k:n)}$ to each side, the desired result follows.

Theorem 2.9 is the more commonly quoted of the two theorems. It is sometimes called Karamata's theorem. Karamata's (1932) proof of this result antedated HLP's proof which appeared in their 1959 book. However HLP had stated the result sans proof in their earlier brief note in the Messenger of Mathematics (in 1929). To them (HLP) the fact that linear combinations of angles (functions of the form $g(x) = (x-a)^+$) were dense in the set of continuous convex functions was intuitively clear. Perhaps they consequently felt it unnecessary to overburden their note with elementary proofs.

Another important class of Schur convex functions are those introduced by Muirhead (1903), subsequently dubbed symmetrical means by HLP (1959). For a given vector $(a_1,\ldots,a_n)$ with each $a_i > 0$ we define the $\underline{x}$'th symmetrical mean of $\underline{a}$ for some set of n real numbers $x_1,\ldots,x_n$ to be

$$[\underline{x}]_{\underline{a}} = [1/n!] \sum_{\pi} a_{\pi(1)}^{x_1} a_{\pi(2)}^{x_2} \cdots a_{\pi(n)}^{x_n} \qquad (2.18)$$

where the sum is over all permutations of the integers $1,2,\ldots,n$. Muirhead (1903) showed that majorization could be verified merely by checking that all symmetrical means were appropriately ordered. He restricted attention to integer valued x_i's, but his proof carries over to the case of more general values for the x_i's, as noted by HLP (1959).

Theorem 2.11 (Muirhead, HLP): $\underline{x} \le_M \underline{y}$ if and only if for every $\underline{a} > \underline{0}$, $[\underline{x}]_{\underline{a}} \le [\underline{y}]_{\underline{a}}$.

__Proof__: A symmetrical mean is evidently symmetric and is easily verified to be convex. From exercise 4 we then conclude that it is Schur convex, i.e., $[\underline{x}]_{\underline{a}} \le [\underline{y}]_{\underline{a}}$ whenever $\underline{x} \le_M \underline{y}$.

Conversely, for a fixed $k < n$, define $\underline{a}^{(k)}(u)$ by $a_1 = a_2 = \ldots = a_k = u$ and $a_{k+1} = \ldots = a_n = 1$. The corresponding symmetric means $[\underline{x}]_{\underline{a}^{(k)}(u)}$ and $[\underline{y}]_{\underline{a}^{(k)}(u)}$ are (generalized) polynomials in u with the indices of their highest powers being respectively $\sum_{i=1}^{k} x_{(i:n)}$ and $\sum_{i=1}^{k} y_{(i:n)}$. In order to have $[\underline{x}]_{\underline{a}^{(k)}(u)} \le [\underline{y}]_{\underline{a}^{(k)}(u)}$ for u large, we must have $\sum_{i=1}^{k} x_{(i:n)} \le \sum_{i=1}^{k} y_{(i:n)}$. Finally, if we set $\underline{a}^{(n)}(u) = (u,u,\ldots,u)$, we can only have $[\underline{x}]_{\underline{a}^{(n)}(u)} \le [\underline{y}]_{\underline{a}^{(n)}(u)}$ for u both large and small if $\sum_{i=1}^{n} x_{(i:n)} = \sum_{i=1}^{n} y_{(i:n)}$.

The proof of the converse given above is, modulo notation changes, exactly that provided by HLP (1959). The symmetric means are of considerable historical interest. They assume a prominent role in both Muirhead's (1903) paper and in HLP's (1959) Chapter 2. Incidently, at the present juncture we may quickly illustrate Muirhead's development of the arithmetic-geometric mean inequality. Both the arithmetic and the geometric mean of $\underline{a} > \underline{0}$ are symmetric means corresponding respectively to the choices $(1,0,\ldots,0)$ and $(\frac{1}{n}, \frac{1}{n}, \ldots, \frac{1}{n})$ respectively for $\underline{x}$. Since evidently $(1,0,0,\ldots,0) \ge_M (\frac{1}{n}, \ldots, \frac{1}{n})$, the arithmetic-geometric mean inequality follows from Theorem 2.11.

Marshall and Olkin (1979, Chapter 4, Section B) describe other classes of Schur convex functions which may be used to determine majorization. The symmetric means and the separable convex functions remain the classical examples.

Exercises

1. Evidently $\underline{x} = (2,2,2) \leq_M (1,2,3) = \underline{y}$. Verify that although $\underline{x}$ may be obtained from $\underline{y}$ by a single Robin Hood operation, it requires a countable number of elementary Robin Hood operations to obtain $\underline{x}$ from $\underline{y}$.
2. Give an example to show that the doubly stochastic matrix P alluded to in Theorem 2.1 is not necessarily unique.
3. Give an example of a discontinuous Schur convex function defined on $\mathbb{R}_n$.
[Hint: If $\sum_{i=1}^{n} x_i = 1$ and $\sum_{i=1}^{n} y_i = 2$, then $g(\underline{x})$ and $g(\underline{y})$ do not have to be related in any way.] [More pathologically, can you construct a non-measurable Schur convex functon on $\mathbb{R}_n$?].
4. Suppose that g is symmetric on A and convex on A. Prove that g is Schur convex in A. [We really only need convexity on sets of the form $\{\underline{x}: \sum_{i=1}^{n} x_i = c\}$.]
5. Prove that separable convex functions on $\mathbb{R}_n$ are Schur convex without assuming differentiability.
6. Verify that the functions defined in (2.11) and (2.12) are indeed Schur convex but not separable convex.
7. Consider the function $g(\underline{x}) = \sum_{i=1}^{n} (x_i - \bar{x})^2$. Is it Schur convex? Is it separable convex?
8. Suppose $g(\underline{x}) = \sum_{i=1}^{n} a_i x_{i:n}$. Supply suitable conditions on the vector $\underline{a}$ to guarantee that g will be Schur convex.
9. If h is convex on I, then $g(\underline{x}) = \sum_{i=1}^{n} h(x_i)$ is Schur convex on I^n.

   Prove the converse, i.e., if $g(\underline{x}) = \sum_{i=1}^{n} h(x_i)$ is Schur convex on I^n, then h is convex on I.
10. (Easy but useful). If g is Schur convex on A and if we define, for $t \in \mathbb{R}$,

$$\chi_t(\underline{x}) = 1 \quad \text{if } g(\underline{x}) \geq t,$$
$$= 0, \quad \text{otherwise,}$$

then $\chi_t(\underline{x})$ is Schur convex on A.

11. If g is a non-decreasing Schur convex function defined on I^n and h is a convex function on $\mathbb{R}$, verify that

$$\phi(\underline{x}) = g\big(h(x_1), h(x_2), \ldots, h(x_n)\big)$$

is Schur convex on I^n.

12. Suppose h is convex on $\mathbb{R}$ and define

$$g(\underline{x}) = \sum_{\pi} e^{\sum_{i=1}^{n} h(c_{\pi(i)} x_i)}$$

where the sum is over all permutations of $(1,\ldots,n)$. Verify that g is Schur convex (e.g., Muirhead's symmetric means). [Marshall and Olkin (1979) point out that the result is also true when the summation is extended only over the k largest of the n! summands.]

13. Suppose that $\sum_{i=1}^{n} x_i = \sum_{i=1}^{n} y_i$ and $x_{i:n}/y_{i:n}$ is a non-increasing function of i. Show that $\underline{x} \leq_M \underline{y}$.

14. Suppose the P is a doubly stochastic matrix. Verify that the matrix $I - P'P$ is positive semi-definite. Use this observation to prove that $g(\underline{x}) = \sum_{i=1}^{n} x_i^2$ is Schur convex.

CHAPTER 3

THE LORENZ ORDER IN THE SPACE OF DISTRIBUTION FUNCTIONS

The graphical measure of inequality proposed by Lorenz (1905) in an income inequality context is intimately related to the concept of majorization. The Lorenz curve, however, can be meaningfully used to compare arbitrary distributions rather than distributions concentrated on n points, as is the case with the majorization partial order. The Lorenz order can, thus, be thought of as a useful generalization of the majorization order. While extending our domain of definitions in one direction, to general rather than discrete distributions, we find it convenient to add a restriction which was not assumed in Chapter 2, a restriction that our distributions be supported on the non-negative reals and have finite expectation. In an income or wealth distribution context the restriction to non-negative incomes is often acceptable. The restriction to distributions with finite means is potentially more troublesome. Any real world (finite) population will have a (sample) distribution with finite mean. However, a commonly used approximation to real world income distributions, the Pareto distribution, only has a finite mean if the relevant shape parameter is suitably restricted. See Arnold (1983) for a detailed discussion of Pareto distributions in the income modelling context. To avoid distorted Lorenz curves (as alluded to in exercise 1 and illustrated in Wold (1935)), we will hold fast to our restriction that all distributions to be discussed will be supported on $\mathbb{R}^+$ and will have finite means. In terms of random variables our restriction is that they be non-negative with finite expectations. We will speak interchangeably of our Lorenz (partial) order as being defined on the class of distributions (supported on $\mathbb{R}^+$ with finite means) or as being defined on the class of integrable non-negative random variables.

First we need to recall Lorenz's original definition of his inequality curve, and then describe a version of it suitable for our purposes, i.e. one which can be used to (partially) order non-negative integrable random variables. After rotating and rescaling Lorenz's diagram, we may describe one of his curves as follows. The Lorenz curve corresponding to a particular population of individuals is a function, say $L(u)$, defined on the interval $[0,1]$ such that for each $u \in [0,1]$, $L(u)$ represents the proportion of the total income of the population which accrues to the poorest $100u$ percent of the population. Associated with such a finite population of n individual incomes is a sample distribution function say $F_n(x)$ where, by definition,

$$F_n(x) = (\# \text{ of individuals with income } \leq x)/n . \tag{3.1}$$

How is Lorenz's curve related to this sample distribution function? After a little thought we realize that something is lacking in Lorenz's original definition. The curve is not well defined for every $u \in [0,1]$, only for $u = 0, \frac{1}{n}, \frac{2}{n}, \ldots, \frac{n-1}{n}, 1$ where n is the size of the population. It is reasonable to complete the curve by linear interpolation, and that is what we shall do. If we denote the ordered individual incomes in the population by $x_{1:n}, x_{2:n}, \ldots, x_{n:n}$, then for $i = 1, \ldots, n$

$$L(\tfrac{i}{n}) = \left(\sum_{j=1}^{i} x_{j:n} \Big/ \sum_{j=1}^{n} x_{j:n}\right) . \tag{3.2}$$

The points $(\frac{i}{n}, L(\frac{i}{n}))$ are then linearly interpolated to complete the corresponding Lorenz curve. Such a curve does approximate the bow shape alluded to by Lorenz. Obvious modifications are required in (3.2) if the x_i's are not all distinct. The Lorenz curve is well defined at 0 and at a number of points equal to the number of distinct values among the x_i's. It is then completed by linear interpolation. Now any distribution function can be approximated arbitrarily closely by discrete distributions. Thus (3.2) must essentially determine a functional on the space of all distribution functions which will associate a "Lorenz curve" with each distribution in a manner consistent with Lorenz's definition of the curve for sample distribution functions. The most convenient mathematical description of that functional is of recent provenance. It was implicitly known and explicitly available in parametric form before, but seems to have not been clearly enunciated until Gastwirth (1971) supplied the following definition.

For any distribution function F we define the corresponding "inverse distribution function" by

$$F^{-1}(y) = \sup\{x: F(x) \leq y\}, \quad 0 < y < 1 . \tag{3.3}$$

With this definition the mean, μ_F, of the distribution (assumed supported on $[0,\infty)$) is given by

$$\mu_F = \int_0^1 F^{-1}(y)\, dy , \tag{3.4}$$

in the sense that the mean exists if and only if the Riemann integral in (3.4) converges. With this definition of F^{-1}, Gastwirth defines the Lorenz curve corresponding to the distribution F by

$$L(u) = [\int_0^u F^{-1}(y)dy]/[\int_0^1 F^{-1}(y)dy], \quad 0 \leq u \leq 1 . \tag{3.5}$$

It is a straightforward matter to verify that the definition (3.5) does indeed coincide with Lorenz's original definition, i.e., (3.2) with linear interpolation, in the case of a sample distribution function corresponding to n numbers (individual incomes), since $F_n^{-1}(y) = x_{i:n}, \frac{i-1}{n} \leq y < \frac{i}{n}$. The form (3.5) is especially useful since it makes transparent several important properties of Lorenz curves. A Lorenz curve is a continuous function on [0,1] with L(0) = 0 and L(1) = 1. It is non-decreasing and differentiable almost everywhere. Convexity of the Lorenz curve is obvious since the function F^{-1} is non-decreasing. Thus the general definition provided by (3.5) does give us "bow-shaped" curves, as promised by Lorenz. A Lorenz curve will always lie below the 45° line joining (0,0) to (1,1) (by convexity). It will coincide with the 45° line in the case of a degenerate distribution. It is evident that the Lorenz curve determines the distribution up to a scale transformation ($L'(u) = cF^{-1}(u)$ a.e. and F^{-1} determines F).

If the Lorenz curve is twice differentiable in some interval say (u_1, u_2), then the corresponding distribution has a finite positive density in the interval $(\mu_F L'(u_1+), \mu_F L'(u_2-))$ where μ_F is defined in (3.4). The density in that interval is given by

$$f(x) = [\mu_F L''(F(x))]^{-1} . \tag{3.6}$$

For an example of a Lorenz curve consider the classical Pareto distribution defined by

$$F(x) = 1 - (x/\sigma)^{-\alpha}, \quad x > \sigma \tag{3.7}$$

where $\sigma > 0$ and $\alpha > 1$ (to ensure the existence of the mean). One finds

$$F^{-1}(u) = \sigma(1-u)^{-1/\alpha}, \quad 0 < u < 1$$

and consequently, from (3.5),

$$L(u) = 1 - (1-u)^{(\alpha-1)/\alpha}, \quad 0 \le u \le 1 . \tag{3.8}$$

Lorenz proposed ordering income distributions by the degree with which the Lorenz curve bow is bent; or more prosaically in terms of nested Lorenz curves. One associates a high level of inequality with a severely bent bow. The case of complete equality corresponds to the unbent bow or 45° line. We can associate an integrable non-negative random variable with any distribution function supported on $[0,\infty)$ with finite mean. We thus can and will discuss a Lorenz partial order on the class of all non-negative integrable random variables rather than on the class of distribution functions. For any non-negative random variable X we will denote its corresponding distribution function by F_X and its corresponding Lorenz curve (i.e. the Lorenz curve corresponding to F_X via (3.5)) by L_X. With this notation, we define the Lorenz partial order by

Definition 3.1: $X \leq_L Y$ (i.e., X does not exhibit more inequality in the Lorenz sense than does Y), if $L_X(u) \geq L_Y(u)$ for every $u \in [0,1]$.

For example, suppose that X and Y have classical Pareto distributions with parameters (α_1,σ_1) and (α_2,σ_2) respectively (refer to equation (3.7)). Assume that $\alpha_1 < \alpha_2$. By referring to the calculated Lorenz curve (3.8), it is evident that $L_X(u) < L_Y(u)$, $\forall\ u \in (0,1)$ and hence $X \geq_L Y$. Thus, in the case of the classical Pareto distribution an increase in the parameter α corresponds to the decrease in inequality as measured by the Lorenz ordering.

Although the Lorenz inequality ordering has achieved a remarkable degree of acceptance, especially in the economic arena, the development of related analytic theory for popular parametric families of densities has been slow. For any finite population there is no problem evaluating the Lorenz curve. For a continuous distribution an analytic expression for the Lorenz curve will rarely be available. This is true because one must first get an analytic expression for the inverse function, and then hope one can integrate it. Fortunately, it is sometimes possible to determine that Lorenz curves are nested without explicitly deriving the Lorenz curves in question (see the example following Theorem 3.4 below, and also the material in Chapter 4).

Another possibility involves use of a parametric representation of the Lorenz curve. Most of the early work on Lorenz curves was done in terms of such a representation. Corresponding to a distribution function F (supported on $[0,\infty)$ with finite mean) we define its first moment distribution function to be

$$F^{(1)}(x) = [\int_0^x y\ dF(y)]/[\int_0^\infty y\ dF(y)]\ . \tag{3.9}$$

A parametric representation of the Lorenz curve is then possible, as follows. The Lorenz curve corresponding to the distribution F is the set of points $(F(x), F^{(1)}(x))$ in the unit square where x ranges from 0 to ∞ completed, if necessary, by linear interpolation. By using this representation a graphical comparison of the Lorenz curves of two distributions F and G is clearly possible. Analytically the comparison involves the quantities $F^{(1)}(x)$ and $G^{(1)}(G^{-1}(F(x)))$. A slight computational gain is observable here in that we only have to invert one of the distribution functions. Exercise 4 illustrates this technique. In this exercise it is verified that if $X \sim \Gamma(1,1)$ and $Y \sim \Gamma(2,1)$, then $X \geq_L Y$. In this example the distribution function for X is readily invertible while that for Y is not. The more general conclusion that if $X \sim \Gamma(k_1,1)$ and $Y \sim \Gamma(k_2,1)$ with $k_1 < k_2$ then, $X \geq_L Y$ is included in Taillie (1981).

A succinct expression for the Lorenz curve can be formed utilizing (3.9). One may write

$$L(u) = F^{(1)}(F^{-1}(u))\ . \tag{3.10}$$

To use this result, closed form expressions for both F^{-1} and $F^{(1)}$ are needed.

The list of parametric families of distributions for which closed form expressions are available for the corresponding Lorenz curve is remarkably short. It includes the family of classical Pareto distributions (equation 3.8 above), distributions uniform on finite intervals, exponential distributions (exercise 9) and arc-sin distributions. An alternative approach is to propose parametric families of Lorenz curves (whose corresponding distribution functions are usually not simple or even available in closed form) to be used in fitting observed Lorenz curves. A simple example is

$$L(u) = 1 - (1 - u^{\beta})^{\alpha}$$

where $\alpha \in (0,1]$ and $\beta \geq 1$. An alternative family is

$$L(u) = [1 - (1 - u)^{\alpha}]^{\beta}$$

where $0 < \alpha \leq 1 \leq \beta$.

The lognormal distribution has a remarkable representation for it Lorenz curve namely,

$$L(u) = \Phi(\Phi^{-1}(u) - \sigma) \tag{3.11}$$

(see exercise 10) where Φ is the standard normal distribution function and σ is the scale parameter for log X. The expression (3.11) can be used to generate other parametric families of Lorenz curves by replacing Φ by some other distribution function. It turns out that a sufficient condition for (3.11) to represent a family of Lorenz curves is that the distribution function Φ be strongly unimodal (Arnold et al (1987)). A judicious non-normal choice for Φ in (3.11) will yield the family of classical Pareto Lorenz curves, (3.8) (see exercise 11). No matter which strongly unimodal Φ is used in (3.11), the resulting family is Lorenz ordered by σ.

A surprisingly good parametric family of Lorenz curves for fitting observed income distributions are the general quadratic curves. In many situations the observed graph $\{(u,L(u)): 0 \leq u \leq 1\}$ is well approximated by a segment of an ellipse. Such Lorenz curves are remarkable in that it is possible to derive closed form (albeit not aesthetically pleasing) expressions for the corresponding distribution and density functions (Arnold and Villaseñor (1984)).

Functionals of the Lorenz curve have been proposed as simple summary measures of inequality. The most popular such measure is the Gini index. The Gini index is conveniently defined as twice the area between the Lorenz curve and the 45° line (which is the Lorenz curve corresponding to an egalitarian distribution in which all individuals have identical incomes). The Pietra index is the maximal vertical deviation between the Lorenz curve and the egalitarian line. A third proposal is an index defined to be simply the length of the Lorenz curve (proposed, for example, by Kakwani (1980)). Some alternative representations of these inequality measures are investigated in the exercises at the end of this chapter.

At this point it is convenient to relate the majorization partial order to the Lorenz order described in this chapter. Majorization is a partial order on n-tuples of real numbers. In the present context we restrict its domain to the non-negative orthant, i.e., sets of n non-negative numbers. Any such set of n numbers can have a (sample) distribution function associated with it (cf. equation (3.1)). If we use Gastwirth's definition of the Lorenz curve (3.5), we may see (using (3.2)) that for two vectors $\underline{x}$, $\underline{y}$ in $\mathbb{R}_n^+$ we have $\underline{x} \leq_M \underline{y}$ if and only if $\sum_{i=1}^{n} x_i = \sum_{i=1}^{n} y_i$ and $X \leq_L Y$ where the random variables X and Y are defined by

$$P(X = x_i) = \frac{1}{n}, \qquad i = 1,2,\dots,n$$
$$P(Y = y_i) = \frac{1}{n}, \qquad i = 1,2,\dots,n . \tag{3.12}$$

Returning to Gastwirth's general definition of the Lorenz curve (3.5), it is immediately apparent that $X \leq_L Y$ if and only if $X/E(X) \leq_L Y/E(Y)$ (we have assumed non-negativity and integrability for our random variables and we tacitly exclude random variables degenerate at 0 so we can divide by expectations). Thus, the Lorenz order actually relates equivalence classes of random variables where two random variables are considered equivalent if one is a scalar multiple of the other. If we use (3.12) as a device to define a partial order on $\mathbb{R}_n^+$, conveniently called the Lorenz order and denoted by $\leq_L$, we see that $\underline{x} \leq_L \underline{y}$ if and only if the normalized vectors $\tilde{\underline{x}}$, $\tilde{\underline{y}}$ satisfy $\tilde{\underline{x}} \leq_M \tilde{\underline{y}}$ (where $\tilde{x}_i = x_i/\sum_{i=1}^{n} x_i$, $\tilde{y}_i = y_i/\sum_{i=1}^{n} y_i$). The relationship between the two orders $\leq_L$ and $\leq_M$ on $\mathbb{R}_n^+$ is evidently intimate but their distinct nature is exemplified by the observation that if $\sum_{i=1}^{n} x_i \neq \sum_{i=1}^{n} y_i$ then $\underline{x}$ and $\underline{y}$ cannot be related by $\leq_M$ but might be related by $\leq_L$.

In Chapter 2, we encountered some remarkable equivalent conditions for majorization in $\mathbb{R}_n$. Specifically consider Theorems 2.1, 2.9 and 2.10. To what extent can we extend these results to deal with the Lorenz partial order on non-negative integrable random variables (which can legitimately be viewed as an extension of majorization)?

Theorems 2.9 and 2.10 extend readily. Thus

Theorem 3.2: Suppose $X \geq 0$, $Y \geq 0$ and $E(X) = E(Y)$. We have $X \leq_L Y$ if and only if $E(h(X)) \leq E(h(Y))$ for every continuous convex function $h: \mathbb{R}^+ \to \mathbb{R}$.

We have the obvious

Corollary 3.2.1. $X \leq_L Y$ iff $E[g(X/E(X))] \leq E[g(Y/E(Y))]$ for every convex continuous g.

or the sometimes more convenient

Corollary 3.2.2. $X \leq_L Y$ iff $E[g(E(Y)X)] \leq E[g(E(X)Y)]$ for every convex continuous g.

Theorem 3.3: Suppose $X \geq 0$, $Y \geq 0$ and $E(X) = E(Y)$. We have $X \leq_L Y$ if and only if $E[(X-c)^+] \leq [E(Y-c)^+]$ for every $c \in \mathbb{R}^+$.

It is not easy to track down explicit proofs of these theorems in the literature. Of course, if X and Y have only n equally likely possible values, the theorems reduce to the earlier proved Theorems 2.9 and 2.10. The general proof then follows easily by a limiting argument. Actually HLP (1929) includes the more general result. To make the correspondence one has to rewrite $E(h(X))$ as $\int_0^1 h(F_X^{-1}(u))du$ and $E(h(Y))$ as $\int_0^1 h(F_Y^{-1}(u))du$. The HLP result involves monotone rearrangements of funtions. But since F_X^{-1} and F_Y^{-1} are already monotone the HLP (1929) theorem is indeed equivalent to Theorem 3.2 provided that, as assumed, $E(X) = \int_0^1 F_X^{-1}(u)du = \int_0^1 F_Y^{-1}(u)du = E(Y)$. See also Marshall and Olkin (1979, p. 15).

Theorem 3.2 and its corollaries suggest that a reasonable summary measure of inequality will be provided by an index of the form $E(g(X/E(X)))$ for any continuous convex g. The choice $g(x) = x^2$ leads to an ordering equivalent to that based on the coefficient of variation (see exercise 20).

Now we turn to the possibility of extending Theorem 2.1 to the more general context of the Lorenz order on integrable non-negative random variables. To this end, let us first look at Theorem 2.1 from a slightly different perspective. Recall that the theorem states that $\underline{x} \geq_M \underline{y}$ if and only if $\underline{y} = P\underline{x}$ for some doubly stochastic matrix P. We may rewrite this as a statement involving random variables X and Y as defined in (3.12). But then what is the role of P? For discussion purposes it is convenient to assume that the coordinates of $\underline{x}$ and of $\underline{y}$ are distinct (i.e. $i \neq i'$ implies $x_i \neq x_i'$ and $y_i \neq y_i'$). On some convenient probability space construct a bivariate random variable (X,Z) where X has possible values $x_1, x_2, \ldots, x_n$ and Z has possible values $1,2,\ldots,n$. Suppose that the joint distribution of (X,Z) is described by:

$$P(Z = i) = 1/n, \quad i = 1,2,\ldots,n$$

and

$$P(X = x_j | Z = i) = p_{ij}, \quad i,j = 1,2,\ldots,n$$

where the p_{ij}'s are elements of a doubly stochatic matrix P. If we denote $E(X|Z=i)$ by y_i, it is really verified that

$$y_i = \sum_{i=1}^{n} p_{ij}x_j \ .$$

Now $P(E(X|Z) = y_i) = P(Z = i) = 1/n$ so that $E(X|Z) \stackrel{d}{=} Y$ (where Y is as defined in (3.12)). What we have shown is that in the context of random variables X and Y each with n equally likely distinct possible values, $X \geq_L Y$ if and only if there exist jointly distributed random variables X', Z' with $X \stackrel{d}{=} X'$ and $Y \stackrel{d}{=} E(X' | Z')$. This theorem is true in much more general settings. Restating a theorem of Strassen (1965) in terms of the Lorenz order we have:

Theorem 3.4: Let X and Y be non-negative integrable random variables with $E(X) = E(Y)$. $Y \leq_L X$ if and only if there exist jointly distributed random variables X', Z' such that $X \stackrel{d}{=} X'$ and $Y \stackrel{d}{=} E(X' | Z')$.

Proof: We will prove only the easy part, referring the reader to Strassen's paper for the more difficult converse. Suppose that $Y = E(X' | Z')$, we claim that $Y \leq_L X'$. Obviously, $E(Y) = E(X')$ so that by Theorem 3.2 it will suffice to verify that $E(h(Y)) \leq E(h(X'))$ for every continuous convex h. This however is true since

$$\begin{aligned} E(h(X')) &= E(E(h(X') | Z')) \\ &\geq E(h(E(X' | Z'))) \qquad \text{(Jensen's inequality)} \\ &= E(h(Y)) \ . \end{aligned}$$

The conditional expectation of X given Z is an averaging of X, and our Theorem 3.4 merely quantifies the plausible statement that, in general, averaging will decrease inequality (as measured by the Lorenz ordering). The reverse operation is known as balayage or sweeping out. We could then phrase our theorem in the form: balayages increase inequality.

Suppose X has an exponential (λ) distribution and Y has a distribution function of the form

$$F_Y(y) = 1 - \left(1 + \frac{y}{\sigma}\right)^{-\alpha}, \quad y > 0 \ , \tag{3.13}$$

where $\alpha > 1$ and $\sigma > 0$. Y can be said to have a Pareto (II) distribution (a translated classical Pareto distribution). We claim that $X \leq_L Y$. One approach would involve direct computation and comparison of the corresponding Lorenz curves. However, a simple observation allows us to draw the conclusion as a consequence of Theorem 3.4.

Suppose that X and Z are independent random variable with $X \sim \Gamma(1,\lambda^{-1})$ and $Z \sim \Gamma(\alpha,1)$. If we define $Y = X/Z$, then by direct computation we find that Y has the Pareto (II) distribution given by (3.13). However, by construction $E(Y|X) = XE(1/Z)$ where X has an exponential (λ) distribution. Thus, $X = E(Y/E(Z^{-1})|X)$, and so by Theorem 3.4, $X \leq_L Y/E(Z^{-1})$, and by the scale invariance of the Lorenz order, $X \leq_L Y$.

Theorem 3.4 would be especially useful if one could derive some algorithm which, for a given pair of random variables X,Y ordered by $Y \leq_L X$, would generate the distribution of the appropriate bivariate random variable (X',Z') referred to in the theorem. Put more bluntly, how does one recognize a balayage? We will return to this problem in Chapter 4 when we discuss inequality attenuating transformations.

Exercises

1. Suppose we were to use definition 3.5 for a random variable which assumes negative values. How will the resulting "Lorenz curve" differ from that usually encountered? Illustrate with the case of a random variable distributed uniformly over the interval (-1,2).
2. Verify (3.4).
3. Let $L(u)$ be a Lorenz curve which for some $u^* \in (0,1)$ satisfies $L(u^*) = u^*$. Prove that $L(u) = u, \forall u \in [0,1]$.
4. Suppose $X \sim \Gamma(1,1)$ and $Y \sim \Gamma(2,1)$. Prove that $X \geq_L Y$. [Hint: Compare $F_Y^{(1)}(y)$ and $F_X^{(1)}(F_X^{-1}(F_Y(y)))$].
5. Provide a careful proof of Theorem 3.2 using the monotone convergence theorem and Theorem 2.9.
6. For any j, the function $g_j(\underline{x}) = \sum_{i=1}^{j} x_{i:n} / \sum_{i=1}^{n} x_{i:n}$ is Schur concave. Verify that $\underline{x} \leq_M \underline{y}$ iff $g_j(\underline{x}) \geq g_j(\underline{y})$, $j = 1,2,\ldots,n$.
7. Verify that convex combinations of Lorenz curves are again Lorenz curves.
8. Let $Z = \mu + \sigma X$ where $\mu \geq 0$ and $\sigma > 0$. Assume $X \geq 0$ and $0 < E(X) < \infty$.

   (a) Let $g(\mu,\sigma,u) = L_Z(u)$. Show that for fixed μ, $g(\mu,\sigma,u)$ is a non-decreasing function of σ. Thus, if $\sigma_1 < \sigma_2$, we have $\mu + \sigma_1 X \geq_L \mu + \sigma_2 X$.

   (b) Assume $\sigma = 1$. Investigate the relationship between the Lorenz curves of X and Z (hint: $F_Z^{-1}(u) = \mu + E(X)L_X'(u)$).
9. (Lorenz curve for exponential distribution.) Suppose X has density $f_X(x) = \lambda e^{-\lambda x}$, $x > 0$. Determine the form of the corresponding Lorenz curve of X.

10. Suppose X has a lognormal distribution, i.e., $\log X \sim N(0,\sigma^2)$. Verify that the Lorenz curve of X has the form $L_X(u) = \Phi(\Phi^{-1}(u) - \sigma)$ where Φ is the standard normal distribution function.

11. Show that (3.8) can be thought of as being a special case of (3.11) for a suitable strongly unimodal distribution function Φ.

12. The Lorenz curve corresponding to a particular random variable X is itself a continuous distribution function with support [0,1]. The mean of this distribution function, i.e., $\int_0^1 u\, dL_X(u)$, can be used as a summary measure of inequality of X. How is this measure related to the Gini index of X?

13. [Uniform record values]. Consider repeated sampling from a uniform distribution on the interval [0,1]. Let $Y_1, Y_2, \ldots$ be the sequence of upper record values. Prove that $Y_i \leq_L Y_{i+1}$, $i = 1,2,\ldots$.

14. A Lorenz curve is symmetric if $L(1 - L(u)) = 1 - u$ for every $u \in (0,1)$. Suppose a random variable X has mean μ and density $f(x)$. Show that its Lorenz curve is symmetric if and only if

$$\frac{f(\mu^2/x)}{f(x)} = \left(\frac{x}{\mu}\right)^3$$

for every x for which $f(x) > 0$.

15. (Gini index). The Gini index is defined by

$$G(X) = 2\int_0^1 (u - L_X(u))\, du .$$

Let X_1, X_2 be i.i.d. copies of X and denote their minimum by $X_{1:2}$. Verify that

$$G(X) = E(|X_1 - X_2|)/2E(X)$$

and

$$G(X) = 1 - [E(X_{1:2})/E(X)] .$$

16. (Pietra Index). The Pietra index is defined by

$$P(X) = \max_{u \in (0,1)} [u - L_X(u)] .$$

Assume F_X is strictly increasing on its support and verify that the maximum is achieved when $u = F_X(E(X))$ and that the Pietra index can be expressed as

$$P(X) = E|X - E(X)|/2E(X) .$$

17. Verify that the Pietra index is twice the area of the largest triangle which can be inscribed between the Lorenz curve and the egalitarian line.

18. Since the length of the Lorenz curve must be in the interval $(\sqrt{2}, 2)$, Kakwani proposed the following index of inequality.

$$K(X) = \frac{\ell_X - \sqrt{2}}{2 - \sqrt{2}}$$

where ℓ_X is the length of the Lorenz curve corresponding to F_X. Verify that

$$\ell_X = \frac{1}{E(X)} E\left(\sqrt{[E(X)]^2 + X^2}\right) .$$

19. Let X have a classical Pareto distribution (eq. (3.7)). Determine the corresponding Gini, Pietra and Kakwani indices (defined in exercise 15-18.

20. The coefficient of variation of X is defined by c.v. (X) = $\sqrt{\text{var}(X)}/E(X)$. Verify that if $X \le_L Y$ then c.v. (X) $\le$ c.v. (Y). Is the converse true?

21. Instead of considering the maximum vertical deviation between the Lorenz curve and the egalitarian line (as Pietra did), one might consider the maximum horizontal deviation. Discuss this summary measure of inequality.

22. Using (3.5), determine the density of a random variable X whose mean is 2 and whose Lorenz curve is given by

$$L(u) = u/(9 - 8u), \quad 0 \leq u \leq 1$$

(Aggarwal and Singh (1984)).

CHAPTER 4

TRANSFORMATIONS AND THEIR EFFECTS

We may most easily motivate the material in the present chapter by setting it in the context of income distributions. Income distributions which exhibit a high degree of inequality (as indicated by their Lorenz curves) are generally considered to be undesirable. Consequently, there are frequent attempts to modify observed income distributions by means of intervention in the economic process. Taxation and welfare programs are obvious examples. Essentially then, we replace the original set of incomes (or, more abstractly, a vector in $\mathbb{R}_n^+$) by some function of the set of incomes. Interest centers on characterizing inequality preserving and inequality attenuating transformations. We will consider both deterministic and stochastic transformations.

Having motivated our task by considering vectors in $\mathbb{R}_n^+$, we make the natural extension suggested by the material in Chapter 3 and consider functions of non-negative integrable random variables. Thus, we seek more insight into two classes of functions mapping $\mathbb{R}^+$ into $\mathbb{R}^+$:

(i) Inequality preserving functions. g is inequality preserving if $X \leq_L Y$ implies $g(X) \leq_L g(Y)$.

(ii) Inequality attenuating functions. g is inequality attenuating if for every non-negative random variable X with $0 < E(X) < \infty$ we have $g(X) \leq_L X$.

An obvious example of an inequality preserving transformation is the function $g(x) = cx$ for some $c > 0$. Are there others? The answer is yes. But there are not very many interesting ones. One may verify that the only inequality preserving trnsformations are those of the following three types:

$$g_{1c}(x) = cx, \quad x \geq 0 \quad \text{for some } c > 0 \;, \tag{4.1}$$

$$g_{2c}(x) = c, \quad x \geq 0 \quad \text{for some } c > 0 \;, \tag{4.2}$$

and

$$\begin{aligned} g_{3c}(x) &= 0, \quad x = 0 \\ &= c, \quad x > 0 \quad \text{for some } c > 0 \;. \end{aligned} \tag{4.3}$$

The classes (4.2) and (4.3) preserve the Lorenz order by mapping the class of non-negative random variables into very restricted classes. As such, they may be considered to be of only academic interest. The proof that (4.1) - (4.3) constitutes a complete enumeration of the inequality preserving transformations involves a rather tiresome enumeration of cases (see Arnold and Villaseñor (1985) for details). Key observations in the arguments are that if $g(0) \neq 0$ then, to preserve inequality, g must be non-decreasing on $(0,\infty)$. Consequently, any inequality preserving function must be measurable. In fact, one can show they must be linear on $(0,\infty)$. Marshall and Olkin (1979, p. 116) show that functions on $\mathbb{R}$ which preserve majorization if measurable must be linear, but in their context, they cannot rule out non-measurable solutions. By restricting attention to functions on $\mathbb{R}^+$ and by asking for preservation of the Lorenz order we are able to avoid such anomalies.

What about inequality attenuating transformations? Sufficient conditions for inequality attenuation have repeatedly been discovered in the literature. In fact, the commonly quoted conditions are essentially necessary and sufficient as we shall now show. Earlier references to the conditions as necesary conditions are to be found in Marshall, Olkin and Proschan (1967), Fellman (1976), Kakwani (1980) and Nygard and Sandstrom (1981). Marshall, Olkin and Proschan (1967) are concerned with *-ordering (see Chapter 6) which implies Lorenz ordering. The Lorenz order is defined on random variables which are non-negative and whose expectations exist and are positive. The positivity requirement is needed to rule out the random variable degenerate at 0. We denote this class of random variables by $\mathscr{L}$. Our theorem can then be stated as follows

Theorem 4.1. Let $g: \mathbb{R}^+ \to \mathbb{R}^+$. The following are equivalent

(i) $g(X) \leq_L X \ \forall X \in \mathscr{L}$

(ii) $g(x) > 0 \ \forall x > 0$, $g(x)$ is non-decreasing on $[0,\infty)$ and $g(x)/x$ is non-increasing on $(0,\infty)$.

Proof: (ii) => (i). Assume g satisfies (ii). Now if $X \in \mathscr{L}$, we need to verify that $g(X) \in \mathscr{L}$. Since $g(x) > 0 \ \forall x > 0$ and $E(X) > 0$, it follows that $E[g(X)] > 0$. Next, since $g(x)$ is non-decreasing on $[0,\infty)$, we have $g(X) \leq g(1)$ when $X \leq 1$. Since $g(x)/x$ is non-increasing on $(0,\infty)$, we have $g(x)/x \leq g(1)/1$ or $g(X) \leq Xg(1)$ when $X \geq 1$. Thus $g(X) \leq (X + 1)g(1)$, and hence $E[g(X)] < \infty$. Thus $g(X) \in \mathscr{L}$. To compare the Lorenz curves of X and $Y = g(X)$, it suffices to consider conveniently chosen random variables X' and Y' with $X \stackrel{d}{=} X'$ and $Y \stackrel{d}{=} Y'$. Let U be a random variable uniformly distributed on the interval (0,1) and let F_X be the distribution function of the random variable X. Define $X' = F_X^{-1}(U)$ and $Y' = g(F_X^{-1}(U))$ (where F_X^{-1} is defined in equation (3.3)). Note that both F_X^{-1} and g are non-decreasing. It follows that

$$\begin{aligned} L_Y(u) - L_X(u) &= L_{Y'}(u) - L_{X'}(u) \\ &= \int_0^u g(F_X^{-1}(v))dv/\int_0^1 g(F_X^{-1}(v))dv \\ &\quad - \int_0^u F_X^{-1}(v)dv/\int_0^1 F_X^{-1}(v)dv \\ &= \int_0^u [g(F_X^{-1}(v)) - F_X^{-1}(v)\,\frac{E(Y')}{E(X')}]\,\frac{dv}{E(Y')} \,. \end{aligned}$$

Since $g(x)/x$ is non-increasing on $(0,\infty)$, the last integrand is first positive then negative as v ranges from 0 to 1. Thus, the integral assumes its smallest value when $u = 1$. Since $L_Y(1) - L_X(1) = 0$, it follows that $L_Y(u) \geq L_X(u)\ \forall\ u \in [0,1]$, i.e. $g(X) = Y \leq_L X$.

$\sim$(ii) => $\sim$(i).

Suppose g is such that there exists $x^* > 0$ with $g(x^*) = 0$. Then consider a random variable X such that $P(X = x^*) = 1$. Obviously $X \in \mathscr{L}$, but $P(g(X) = 0) = 1$, so $g(X) \notin \mathscr{L}$, and thus $g(X) \nleq_L X$.

Suppose $g(x) > 0$ for $x > 0$ but g is not non-decreasing on $[0,\infty)$. Thus, there exist x and y with $0 \leq x < y$ and $g(y) < g(x)$. Consider a random variable X such that $P(X = x) = p$, $P(X = y) = 1 - p$. There are two cases to consider.

Case 1. $x = 0$, $g(y) > 0$. Here, see exercise 3, $g(X) \nleq_L X$ provided $p < (g(0) - g(y))/(2g(0) - g(y))$.

Case 2. $x > 0$, $g(y) > 0$. Here, see exercise 4, $g(X) \nleq_L X$ provided $p > (\frac{y}{x} - 1)/(\frac{g(x)}{g(y)} + \frac{y}{x} - 2)$.

Finally, suppose g is non-decreasing, $g(x) > 0$ for $x > 0$ and $g(x)/x$ is not non-increasing on $(0,\infty)$. Thus, there exist x and y such that $0 < x < y$ with $0 < g(x)/x < g(y)/y$. Let X be a random variable defined by $P(X = x) = P(X = y) = 1/2$. One finds $L_{g(X)}(1/2) < L_X(1/2)$, so that $g(X) \nleq_L X$.

The above theorem has an attractive interpretation in terms of taxation policies. If we think of X as representing the distribution of income before taxes and g(X) as representing income after taxes, then in order for our taxation policy g to be guaranteed to reduce inequality (for any pre-tax income distribution), g must satisfy conditions (ii) of Theorem 4.1. It must satisfy $g(x) > 0\ \forall\ x > 0$, i.e., everyone with some income before taxes should still have some money left after taxes. It must be monotone, i.e., if Sally earned more than Joe before taxes, her after tax income must also be more than Joe's after tax income. Finally, we must have $g(x)/x \downarrow$. But this just says that it must be a "progressive" tax which takes proportionally more from the rich than it does from the poor. Most taxation policies do satisfy these conditions and are, thus, inequality attenuating.

Another way to change income distributions is to mandate salary increases which depend on the current level of salary. Thus, if any individual with salary x is given a 100 $\gamma(x)$% increase in salary, his new salary will be $g(x) = x(1 + \gamma(x))$. We can refer to Theorem 4.1 in order to determine conditions on γ under which such a policy will reduce inequality. We assume $\gamma(x) \geq 0$ (i.e., we are indeed speaking of salary increases and not cuts). In order to have $g(x)/x \downarrow$, we must have $\gamma(x) \downarrow$ (i.e. bigger precentage increases go to the poorer individuals). Finally, to have $g(x)\uparrow$, we need $x[1+\gamma(x)]\uparrow$. This means that γ cannot decrease too fast. For example, if we assume differentiability of γ (note that up to this point we have studiously avoided putting smoothness conditions on g or γ), then to have $g\uparrow$ we need

$$\gamma'(x) \geq - [1 + \gamma(x)]/x .$$

Although most taxation policies in vogue do qualify as being inequality attenuating, such is not always the case for policies dealing with salary increases. A not uncommon policy for salary increases which is not necessarily inequality attenuating is of the following type. All employees receiving less than $10,000 will be given a 15% increase, all employees earning between $10,000 and $20,000 will be given a 12% increase, and all others will be given a 10% increase. Although this is a policy involving generous increases, one can expect to hear from disgruntled employees whose previous salary was $10,050.

Analogous arguments to those used in Theorem 4.1 allow one to characterize inequality accentuating transformations. One finds

Theorem 4.2. Let $g: \mathbb{R}^+ \to \mathbb{R}^+$. The following are equivalent

(i) $X \leq_L g(X) \ \forall X \in \mathscr{L}$

(ii) $g(x) > 0 \;\forall\; x > 0$, $g(x)$ is non-decreasing on $[0,\infty)$ and $g(x)/x$ is non-decreasing on $(0,\infty)$.

What happens when we allow our transformation g to have random components? Two basic insights can guide us here. First, the introduction of extraneous randomness or noise should increase inequality and, second, averaging should decrease inequality. The second of these insights is illustrated by Theorem 3.4 which includes the observation that if $Y = E(X|Z)$ (or if $Y \stackrel{d}{=} E(X|Z)$), then $Y \leq_L X$. Thus, if Y can be identified as a conditional expectation of X given some random variable Z, then $Y \leq_L X$. If Y is a conditional expectation of X, then X is said to be a balayage, dilation or sweeping out of Y. The first insight, regarding the introduction of noise, is illustrated in the misreported income example following equation (4.5) below. But, as a caveat, see exercise 9.

We may rephrase our question about possibly random transformations which attenuate or accentuate inequality as follows. Let X be a non-negative random variable, and suppose that

$$Y = \psi(X,Z) \geq 0 \tag{4.4}$$

where Z is random and ψ is deterministic. Under what conditions on ψ, Z and on the joint distribution of Z and X can we conclude that necessarily $Y \leq_L X$ (attenuation) or $Y \geq_L X$ (accentuation)? We can see immediately that inequality accentuation rather than attenuation is most likely to result from transformations of the form (4.4). This is because such transformations usually involve additional randomness or noise. For example, if X is degenerate, then Y, in (4.4), is typically not degenerate so $Y \geq_L X$. A simple example in which inequality attenuation occurs is the following. If $Z \equiv X$ and $\psi(x,z) = g\big((x+z)/2\big)$ where g satisfies conditions (ii) of Theorem 4.1, then $Y \leq_L X$, for any $X \in \mathscr{L}$.

Let us consider a non-trivial example in which inequality accentuation obtains. Several authors have considered the following model for misreporting of income (on tax returns, for example)

$$Y = UX \ . \tag{4.5}$$

Here Y represents reported income, X is true income, and U is the misreporting factor. A common assumption is that U and X are independent (non-negative random variables). It follows immediately that $E(Y|X) = E(U)X$, and so, by Theorem 3.4, $Y \geq_L E(U)X$, whence $Y \geq_L X$. Misreporting increases inequality. In fact we do not even need to assume U and X are independent. It suffices that $E(U|X) = c$.

In practice, it seems reasonable to assume that, in (4.5), $U \leq 1$ (i.e. people underreport their income on their tax returns). A progressive tax (in the sense of Theorem 4.1), of course, is applied to reported income (not actual income). Does it still attenuate inequality? Now, post tax income is of the form

$$Y = (1 - U)X + g(UX) \ . \tag{4.6}$$

Here $Y|U = u$ is a non-random function of X which satisfies conditions (ii) of Theorem 4.1. Thus, $Y|U = u \leq_L X$ for every $u \in [0,1]$. It is tempting to conclude that the results still holds unconditionally, i.e., $Y \leq_L X$. Such an argument is a snare, however. We cannot, in general, expect to have $Y \leq_L X$ when X and Y are related by (4.6). The case of a degenerate X again provides a fly in the ointment. For even though X is degenerate, say $X \equiv 1$, Y is decidedly not degenerate, and so $Y \not\leq_L X$. Continuity arguments allow us to conclude that non-degenerate counterexamples must also exist. Consequently, we cannot be sure that a progressive tax on reported income will necessarily attenuate the inequality of actual income, which is sad but true and, retrospectively, quite obvious.

It is possible to identify random transformations which necessarily accentuate inequality. For example, if U and X are independent non-negative random variables, and if g satisfies condition (ii) of Theorem 4.2, then $Y = g(UX) \geq_L X$. This is a direct consequence of Theorem 4.2, the fact that misreporting accentuates inequality and the fact that the Lorenz order is transitive. We have in this situation $X \leq_L UX \leq_L g(UX)$. A slight generalization of this observation is provided by the following theorem.

Theorem 4.3. Suppose $g: \mathbb{R}_2^+ \to \mathbb{R}^+$ is such that $g(z,x)/x$ is non-decreasing in x for every z, and $g(z,x)$ is non-decreasing in x for every z. Assume that X and Z are independent non-negative random variables with $X \in \mathscr{L}$ and $g(Z,X) \in \mathscr{L}$. It follows that $X \leq_L g(Z,X)$.

Proof: Exercise 6.

In Theorem 4.3 in order to have $X \leq_L g(Z,X)$, we really only require that $g(z,x)/x$ and $g(z,x)$ be non-decreasing in x as x ranges over the set of possible values of X, and we only require this to hold for any z that is a possible value of Z (x is a possible value of X if for every $\varepsilon > 0$, $P(x - \varepsilon < X < x + \varepsilon) > 0$). Similar "extensions" of Theorem 4.1 and 4.2 are possible. For example, in the setting of Theorem 4.1, we have $g(X) \leq_L X$ provided conditions (ii) hold as x ranges over the set of possible values of X.

In the misreported income scenario discussed above, instead of observing X, we observed a transformed version of X. In many scientific fields, the random variable X of interest is also not observed. What is observed is not a transformation of X but, rather, a weighted version of X. The basic reference is Rao (1965). Mahfoud and Patil (1982) provide a more current survey of the area. Instead of observing random variables with density $f(x)$, because of the method of ascertainment (the way the data are collected), we actually observe random variables with a density proportional to $g(x)f(x)$. The function $g(x)$ is the weighting function. The special case $g(x) = x$, called size biased sampling, occurs when bigger units are more likely to be sampled than small ones. How do such weightings affect inequality as measured by the Lorenz order?

Suppose that $X \in \mathscr{L}$ and that g is a suitable measurable non-negative function. The g-weighted version of X, denoted X_g is defined to be a random variable such that

$$P(X_g \leq x) = \int_0^x g(y)dF_X(y)/E[g(X)] \tag{4.7}$$

provided $0 < E(g(X)) < \infty$. Note that if $X \in \mathscr{L}$ then in order to have $X_g \in \mathscr{L}$ we require both $0 < E(g(X)) < \infty$ and $0 < E(Xg(X)) < \infty$.

Inequality preserving weightings will correspond to functions g for which $X \leq_L Y \Rightarrow X_g \leq_L Y_g$. Obviously a homogeneous function of the form $g(x) \equiv c > 0$ will preserve inequality. Using the basic Lemmas described in exercises 1 and 2, it is not difficult to verify that there is very little scope for variation from homogeneity. In fact (Arnold (1986)) the only inequality preserving weightings are of the form

$$g(0) = \alpha$$

$$g(x) = \beta, \quad x > 0$$

where $\alpha \geq \beta > 0$ (the first step in the proof is exercise 14).

In a similar fashion we can seek a "weighting" version of Theorem 4.1. Again the basic lemmas of exercises 1 and 2 are helpful. Very few weightings are inequality attenuating. One may verify (Arnold (1986)) that $X_g \leq_L X$ for every $X \in \mathscr{L}$ if and only if

$$g(x) = \alpha, \quad x = 0$$

$$= \beta, \quad x > 0$$

where $\beta > 0$ and $0 \leq \alpha \leq \beta$ (the first step in the proof is exercise 15).

Exercises

1. Suppose $0 < x_1 < x_2$ and that random variables X and Y are defined by

$$P(X = x_1) = p, \quad P(X = x_2) = 1 - p,$$

$$P(Y = x_1) = p', \quad P(Y = x_2) = 1 - p'.$$

Show that X and Y are not Lorenz ordered except in the trivial cases when $p = p'$, $pp' = 0$ or $(1-p)(1-p') = 0$.

2. Suppose $0 < x$ and that random variables X and Y are defined by

$$P(X = 0) = p, \quad P(X = x) = 1 - p,$$

$$P(Y = 0) = p', \quad P(Y = x) = 1 - p'.$$

Show that $p \leq p' \Rightarrow X \leq_L Y$.

3. Assume g is such that there exists $y > 0$ with $0 < g(y) < g(0)$. Let X be a random variable such that $P(X=0) = p$, $P(X=y) = 1-p$. Show that $g(X) \not\leq_L X$ for small values of p (specifically, for $p < [g(0) - g(y)]/[2g(0) - g(y)]$).

4. Assume g is such that there exist x and y with $0 < x < y$ and $0 < g(y) < g(x)$. Let X be a random variable such that $P(X=x) = p$, $P(X=y) = 1-p$. Show that $g(X) \not\leq_L X$ for large values of p (specifically, for $p > [\frac{y}{x} - 1]/[\frac{g(x)}{g(y)} + \frac{y}{x} - 2]$).

5. (Inequality attenuation in the sense of majorization.) Show that g satisfies conditions (ii) of Theorem 4.1 if and only if for any n and for any $\underline{x} \in \mathbb{R}_n^+$ we have

$$\left(\frac{g(x_1)}{\Sigma g(x_i)}, \ldots, \frac{g(x_n)}{\Sigma g(x_n)}\right) \leq_M \left(\frac{x_1}{\Sigma x_i}, \ldots, \frac{x_n}{\Sigma x_i}\right). \qquad (*)$$

[Note that it is possible to have a function satify (*) for a fixed n without conditions (ii) of Theorem 4.1 being satisfied. For example with n=2, consider $g(0) = 2$, $g(x) = 1$, $x \neq 0$.]

6. Prove Theorem 4.3.

7. Let $X \in \mathscr{L}$ and define $Y = \mu_1 + \sigma_1 X$ and $Z = \mu_2 + \sigma_2 X$ where $\mu_1, \mu_2 \geq 0$ and $\sigma_1, \sigma_2 > 0$. Under what circumstances can we claim

$Y \le_L Z$? This result, when X is a classical Pareto random variable, is discussed in Samuelson (1965). The Lorenz curves of Y and Z were discussed in Chapter 3, exercise 8. Here we can use Theorems 4.1 and 4.2.

8. Suppose $X \le_L Y$, $a > 0$, $b \ge 0$ and $E(X) = E(Y)$. Show that $aX + b \le_L aY + b$. What happens if $E(X) \neq E(Y)$?

9. Does the addition of noise increase inequality? Suppose X, $U \in \mathscr{L}$. Can we conclude that $X + U \ge_L X$? Can we conclude $X + U \le_L X$?

10. Suppose X is a random variable with finite α'th and β'th moments. Prove that $X^\alpha \le_L X^\beta$ if and only if $\alpha \le \beta$.

11. Supply an example in which $X \le_L Y$ yet $X + 1 \not\le_L Y + 1$.

12. (Deterministic underreporting). Suppose that $U = \frac{1}{2}$ with probability one and that g is an inequality attenuating transformation. Can we conclude that for any $X \in \mathscr{L}$ we have

$$Y = (1 - U)X + g(XU) \le_L X?$$

13. (Strong Lorenz Order). We write $X <_L Y$ if $X \le_L Y$ and $Y \not\le_L X$. State and prove a strong Lorenz order version of Theorem 4.1, i.e. give necessary and sufficient conditions on g to ensure that $g(X) <_L X$ for every non-degenerate X in $\mathscr{L}$.

14. Suppose that for some $x_1, x_2 > 0$ we have $g(0) = \alpha$, $g(x_1) = \beta$ and $g(x_2) = \gamma$ where $\gamma > \beta$. Show that such a weighting g does not preserve the Lorenz order. (Hint: consider two random variables $X \le_L Y$ where

$$P(X = 0) = P(X = x_1) = \frac{1}{2}$$

and

$$P(Y = 0) = \frac{1}{2} + \varepsilon$$

$$P(Y = x_2) = \frac{1}{2} - \varepsilon$$

in which $\varepsilon = (\gamma - \beta)/4(\gamma + \beta)$).

15. Suppose that for some $0 < x_1 < x_2$ we have $g(x_1) = \gamma_1 \neq \gamma_2 = g(x_2)$. Show that such a weighting g does not attenuate inequality. (Hint: consider a random variable X such that $P(X = x_1) = P(X = x_2) = \frac{1}{2}$).

CHAPTER 5

MULTIVARIATE AND STOCHASTIC MAJORIZATION

5.1. Multivariate majorization

The temptation to seek multivariate generalizations of majorization and the Lorenz order is strong, and has not been resisted. In an income setting it is reasonable to consider income from several sources or income in incommensurable units. In fact, the idea that income can be measured undimensionally is perhaps the radical point of view, and interest should center on multivariate measures of income. Let us first consider various possible multivariate generalizations of majorization.

Let $\mathbb{R}^+_{n\times m}$ be the set of all $n\times m$ matrices with non-negative real elements. We want to define m-dimensional majorization to be a partial order on $\mathbb{R}^+_{n\times m}$ in such a way as to reduce to ordinary majorization when $m = 1$. One possible definition is to require column by column majorization. For any matrix $X \in \mathbb{R}^+_{n\times m}$ we let $X^{(j)}$ $(j=1,2,\ldots,m)$ denote the j'th column. With this notation we have the following definition.

Definition 5.1: Let $X,Y \in \mathbb{R}^+_{n\times m}$. We say that X is marginally majorized by Y if, for every $j=1,2,\ldots,m$, $X^{(j)} \leq_M Y^{(j)}$, and we write $X \leq_{MM} Y$.

Now from the Hardy, Littlewood and Polya theorem (Theorem 2.1) we know that if $X \leq_{MM} Y$ then there exist doubly stochastic matrices $P_1,P_2,\ldots,P_m$ such that $X^{(j)} = P_j Y^{(j)}$. Of course the P_j's may well be different. If the same choice of doubly stochastic matrix works for every j, we may well have a stronger but perhaps more interesting partial order. It is this partial order that is dubbed majorization in Marshall and Olkin. To capture the spirit of the definition, we will call the relation uniform majorization. We thus have

<u>Definition 5.2</u>: Let $X,Y \in \mathbb{R}^+_{n\times m}$. We say that X is uniformly majorized by Y if there exists a doubly stochastic matrix P such that $X = PY$, and we write $X \le_{UM} Y$.

How are marginal and uniform majorization related? Obviously $X \le_{UM} Y \Rightarrow X \le_{MM} Y$, for one can set $P_j = P$ for each j. It is not difficult to verify that the converse fails i.e., in general $X \le_{MM} Y \not\Rightarrow X \le_{UM} Y$ (exercise 1).

Life in higher dimensions is invariably a little more complicated than a one dimensional existence. This is exemplified by the fact that Robin Hood loses some of his prominence in higher dimensional versions of majorization.

A Robin Hood operation (refer to Chapter 2) involves a transfer of riches from a relatively richer individual to a relatively poorer individual in the population. It is equivalent to multiplication by a doubly stochastic matrix of the form $A = (a_{ij})$ where for some k_1, k_2 and some $\lambda \in [0,1]$

$$\begin{aligned} a_{k_1,k_1} &= 1-\lambda, & a_{k_1,k_2} &= \lambda\ , \\ a_{k_2,k_1} &= \lambda, & a_{k_2,k_2} &= 1-\lambda\ , \\ a_{ij} &= \delta_{ij}, & &\text{otherwise}\ . \end{aligned} \tag{5.1}$$

Matrices satisfying (5.1) will be called Robin Hood matrices. We may then formulate

<u>Definition 5.3</u>: Let $X,Y \in \mathbb{R}^+_{n\times m}$. We say that X is majorized in the Robin Hood sense by Y, if there exists a finite set of Robin Hood matrices (of the form (5.1)) $A_1, A_2, \ldots, A_M$ such that $X = A_1 A_2 \cdots A_M Y$, and we write $X \le_{RH} Y$.

Now from Chapter 2, we know that majorization can be characterized in terms of Robin Hood operations and relabelings. In our definition (5.1) we allowed λ to be greater than 1/2, so that the class includes elementary permutation matrices. However, notice that in definition 5.3, when a permutation matrix appears among the A_j's, it permutes all the columns of Y, i.e., the same permutation is applied to all the columns. We have no guarantee that this will allow us to duplicate every doubly stochastic matrix as a finite product of Robin Hood matrices. Thus, we claim that uniform and Robin Hood majorization are different concepts, since provided $n \geq 3$, there exist doubly stochastic matrices which are <u>not</u> finite products of Robin Hood matrices (in the sense of (5.1)). This may seem paradoxical, since in one dimension we know that Robin Hood majorization and majorization do coincide. The following example provided by Marshall and Olkin (1979) shows that the concepts must be distinguished in higher dimensions.

<u>Example</u>: Suppose

$$X = \begin{bmatrix} 1 & 3 \\ 1/2 & 4 \\ 1/2 & 5 \end{bmatrix} \quad \text{and} \quad Y = \begin{bmatrix} 1 & 2 \\ 1 & 4 \\ 0 & 6 \end{bmatrix}$$

then $X = PY$ where

$$P = (1/2)\begin{bmatrix} 1 & 1 & 0 \\ 1 & 0 & 1 \\ 0 & 1 & 1 \end{bmatrix} \qquad (5.2)$$

and for no other choice of P is $X = PY$. However, this particular doubly stochastic matrix is not expressible as a finite product of Robin Hood matrices (exercise 10). Thus $X \leq_{UM} Y$, but $X \not\leq_{RH} Y$.

There is a fourth possible generalization of majorization to higher dimensions. Recall that we were able to characterize majorization in terms of continuous convex functions on $\mathbb{R}^+$. The definition of convexity is readily extended to functions on $\mathbb{R}_m^+$, and we may formulate

<u>Definition 5.4</u>: Let $X,Y \in \mathbb{R}_{n\times m}$. We say that X is convexly majorized by Y if for every $h: \mathbb{R}_m^+ \to \mathbb{R}^+$ that is continuous and convex we have $\sum_{i=1}^{n} h(X_{(i)}) \leq \sum_{i=1}^{n} h(Y_{(i)})$ (here $X_{(i)}$ is the i'th row of X), and we write $X \leq_{CM} Y$.

It is a simple consequence of Jensen's inequality that $X \leq_{UM} Y \Rightarrow X \leq_{CM} Y$ (exercise 2). In dimensions higher than 1 (i.e., $m \geq 2$) it is an open question whether the converse is true.

The four possible generalizations thus far introduced (marginal, uniform, Robin Hood and convex) are presumably all distinct (for $m \geq 2$) (unless convex and uniform majorization can be proved to coincide), but no one of the four is compelling. Convex majorization has the advantage of extending readily to cover the case of general non-negative m-dimensional random variables. Before considering this generalization, we mention two other possible versions of multivariate majorization. Both involve efforts to define the concept in terms of the better understood concept of univariate majorization. First, consider a definition proposed by Marshall and Olkin (1979).

Definition 5.5: Let $X, Y \in \mathbb{R}^+_{n\times m}$. We say that X is Marshall-Olkin majorized by Y, if $X\underline{a} \leq_M Y\underline{a}$ for every vector $\underline{a} \in \mathbb{R}^+_m$ with $a_i \geq 0$ $i=1,2,\dots,m$ and $\sum_{i=1}^{m} a_i = 1$, and we write $X \leq_{MO} Y$.

An economic interpretation of Marshall-Olkin majorization is possible. Suppose the rows of X represent m-dimensional income vectors (one for each of n individuals in the population). In particular x_{ij} represents income in currency j accruing to individual i. Suppose now that all the incomes are converted into dollars at rates $b_1, b_2, \dots, b_m$. The resulting vector of incomes in dollars is then $X\underline{b}$. Majorization of $X\underline{b}$ by $Y\underline{b}$ is unaffected by normalizing the b_i's so that $\sum_{i=1}^{m} b_i = 1$. It is then evident that $X \leq_{MO} Y$ if the X incomes are majorized by the Y incomes under all possible exchange rates (i.e. if $X\underline{b} \leq_M Y\underline{b}$, $\underline{b} \geq \underline{0}$, $\underline{b} \not\equiv \underline{0}$).

A different approach to the problem of reducing dimension is provided by

Definition 5.6: Let $X, Y \in \mathbb{R}^+_{n\times m}$ and let $g: \mathbb{R}^+_m \to \mathbb{R}^+$. We say that X is g-majorized by Y, if $(g(X_{(1)}), \dots, g(X_{(n)})) \leq_M (g(Y_{(1)}), \dots, g(Y_{(n)}))$, and we write $X \leq_{gM} Y$.

Plausible choices for g in definition 5.6 include, $g(\underline{x}) = \sum_{i=1}^{m} x_i$, $g(\underline{x}) = \sqrt{\sum_{i=1}^{m} x_i^2}$ and $g(\underline{x}) = \max_i x_i$. In an income setting where the x_i's represent incomes of m different types, $g(\underline{x})$ can be interpreted as the utility of the income vector $(x_1,\ldots,x_m)$. If we can agree on a suitable choice for g, then we are enabled to replace an m dimensional income vector by a suitable one dimensional utility. Both Marshall-Olkin and g-majorization extend readily to the case of general non-negative m-dimensional random variables to which we now turn.

Let $\mathscr{L}^{(m)}$ denote the class of m-dimensional random variables whose coordinate random variables are members of $\mathscr{L}$, i.e., are non-negative random variables with positive finite expectations. We consider several generalizations of the Lorenz order to this m-dimensional setting. In the following X_i (respectively Y_i) denotes the i'th coordinate random variable of $\underline{X}$ (respectively $\underline{Y}$), $i=1,2,\ldots,m$.

Definition 5.7: Let $\underline{X},\underline{Y} \in \mathscr{L}^{(m)}$. We will say that $\underline{X}$ is marginally Lorenz dominated by $\underline{Y}$, if for each $i=1,2,\ldots,m$, $X_i \le_L Y_i$, and we write $\underline{X} \le_{ML} \underline{Y}$.

Definition 5.8: Let $\underline{X},\underline{Y} \in \mathscr{L}^{(m)}$. We will say that $\underline{X}$ is Lorenz dominated by $\underline{Y}$, if for every continuous convex function $h\colon \mathbb{R}_m^+ \to \mathbb{R}^+$ we have

$$E\left(h\left(\frac{X_1}{E(X_1)},\ldots,\frac{X_m}{E(X_m)}\right)\right) \le E\left(h\left(\frac{Y_1}{E(Y_1)},\ldots,\frac{Y_m}{E(Y_m)}\right)\right),$$

and we write $\underline{X} \le_L Y$.

Definition 5.9: Let $\underline{X},\underline{Y} \in \mathscr{L}^{(m)}$. We say that $\underline{X}$ is Marshall-Olkin Lorenz dominated by $\underline{Y}$, if for every $\underline{\lambda}$ such that $\lambda_i \ge 0$, $i=1,2,\ldots,m$ and $\sum_{i=1}^{m} \lambda_i = 1$, we have $\sum_{i=1}^{m} \lambda_i X_i \le_L \sum_{i=1}^{m} \lambda_i Y_i$, and we write $\underline{X} \le_{MOL} \underline{Y}$.

Definition 5.10: Let $\underline{X},\underline{Y} \in \mathscr{L}^{(m)}$ and let $g\colon \mathbb{R}_m^+ \to \mathbb{R}^+$. We say that $\underline{X}$ is g-Lorenz dominated by $\underline{Y}$, if $g(\underline{X}) \le_L g(\underline{Y})$, and we write $\underline{X} \le_{gL} \underline{Y}$.

In definition 5.10, g can be interpreted as a utility function (as in the case of g-majorization). In order to have $E(g(\underline{X})) < \infty$ for every $\underline{X} \in \mathscr{L}^{(m)}$, g should be a bounded function. If g is not bounded one might be tempted to replace it by a new bounded utility such as $g^* = g/(1 + g)$. The partial orders $\leq_{gL}$ and $\leq_{g^*L}$ will however not be equivalent (in the light of the results of Chapter 4 on inequality preserving transformations).

Strassen's theorem (Theorem 3.4) carries over to the multivariate setting. Thus, if $E(\underline{X}) = E(\underline{Y})$, then $\underline{X} \leq_L \underline{Y}$ iff there exist jointly distributed random variables $\underline{Y}'$ and Z' such that $\underline{Y} \stackrel{d}{=} \underline{Y}'$ and $\underline{X} \stackrel{d}{=} E(\underline{Y}'|Z')$. See exercises 4 and 5 for applications of this result. Exercise 6 addresses the relationships between three partial orders that have been defined on $\mathscr{L}^{(m)}$.

Of course, the most natural approach to generalizing the Lorenz order on $\mathscr{L}$ to cover higher dimensional cases would be to first generalize the concept of the Lorenz curve to cover m-dimensional random variables. There is a fundamental difficulty in this approach, however. Using the Gastwirth definition, (3.5), the Lorenz curve involves the inverse distribution or quantile function. In higher dimensions this is more troublesome. Even in the case of an m-dimensional distribution function which is strictly monotone in all its arguments, the equation $F_{\underline{X}}(\underline{x}) = p$ defines an (m-1)-dimensional region. Two candidate Lorenz surfaces have been proposed for bivariate random variables (i.e. the case m = 2). Taguchi (1972a,b) suggests the function L(s,t) defined implicitly by

$$s = \int_0^{x_1} \int_0^{x_2} f_{X_1,X_2}(\xi,\eta)d\xi d\eta ,$$

$$t = \int_0^{x_1} \int_0^{x_2} \xi f_{X_1,X_2}(\xi,\eta)d\xi d\eta/E(X_1) , \qquad (5.3)$$

$$L(s,t) = \int_0^{x_1} \int_0^{x_2} \eta f_{X_1,X_2}(\xi,\eta)d\xi d\eta/E(X_2) .$$

Arnold (1983) proposed the function L(s,t) defined implicitly by

$$s = \int_0^{x_1} f_{X_1}(\xi)d\xi ,$$

$$t = \int_0^{x_2} f_{X_2}(\eta)d\eta , \qquad (5.4)$$

$$L(s,t) = \int_0^{x_1} \int_0^{x_2} \xi\eta f_{X_1,X_2}(\xi,\eta)d\xi d\eta / E(X_1X_2)$$

It is easy to visualize how to generalize the latter definition to the case of m dimensional random variables (m > 2). The Taguchi definition does not treat the coordinate random variables in a symmetric fashion, and a natural extension is difficult to envision. One nice feature enjoyed by the surface defined by (5.4) is that in the case of independence, the bivariate Lorenz surface reduces to the product of the marginal Lorenz curves (exercise 7). One could, of course, define a "Lorenz curve" ordering on $\mathscr{L}^{(m)}$ by saying that $\underline{X}$ is more unequal in the Lorenz curve sense than $\underline{Y}$ if $L_{\underline{X}}(\underline{u}) \le L_{\underline{Y}}(\underline{u})$ for all $\underline{u}$, where the surfaces $L_{\underline{X}}(\underline{u})$, $L_{\underline{Y}}(\underline{u})$ are defined by an m-dimensional version of (5.4). The relation of this partial order to those of definitions 5.7 - 5.10 has not been explored.

Summary measures of inequality for multivariate distributions would most naturally be defined with respect to the "Lorenz order" on $\mathscr{L}^{(m)}$, if we could decide on which Lorenz order to use. If we use definition 5.8, then one could justify (as in the one dimensional case described in Chapter 3), use of any measure based on a convex continuous function h. Thus, we could use as a measure of inequality for a specific choice of h, the quantity

$$E\left(h\left(\frac{X_1}{E(X_1)}, \ldots, \frac{X_m}{E(X_m)}\right)\right) .$$

If we could agree on a suitable definition of the Lorenz surface for m-dimensional random variables, we could measure inequality by the (m+1)-dimensional volume between the Lorenz surface of a given distribution and the Lorenz surface of a degenerate distribution (in direct analogy to one interpretation of the Gini index in the case m = 1). Alternatively, (mimicking Kakwani) we could use the m-dimensional volume of the Lorenz surface as an inequality measure.

The Gini index has yet another representation which extends readily to higher dimensions. The Gini index corresponding to a random variable X may be defined to be

$$G(X) = \frac{1}{2} E \left| \frac{X^{(1)} - X^{(2)}}{E(X)} \right| \tag{5.5}$$

where $X^{(1)}$ and $X^{(2)}$ are independent copies of X. So it is proportional to the expected distance between independent normalized copies of X. In m dimensions we can use the same verbal definition to obtain

$$G^{(m)}(\underline{X}) = c_m E \left(\sqrt{\sum_{i=1}^{m} \left[\frac{X_i^{(1)} - X_i^{(2)}}{E(X_i)}\right]^2} \right) \tag{5.6}$$

where $\underline{X}^{(1)}$ and $\underline{X}^{(2)}$ are independent copies of $\underline{X}$. Of course, Euclidean distance is not sacred and other measures of distance could be used in (5.6). The mysterious constant c_m in (5.6) is discussed in exercise 9.

5.2. Stochastic majorization

We return to the original setting for majorization, i.e., vectors in $\mathbb{R}_n$. However, we now consider random variables which take on values in $\mathbb{R}_n$. If $\underline{X}$ and $\underline{Y}$ are n-dimensional random variables, then for given realizations of $\underline{X}$ and $\underline{Y}$, say $\underline{x}$ and $\underline{y}$, we may or may not have $\underline{x} \leq_M \underline{y}$. If we have such a relation for every realization of $(\underline{X},\underline{Y})$, then we have a very strong version of stochastic majorization holding between $\underline{X}$ and $\underline{Y}$. However, we may be interested in weaker versions. Certainly, for most purposes $P(\underline{X} \leq_M \underline{Y}) = 1$ would be more than adequate. We will content ourselves with the requirement that there be versions of $\underline{X}$ and $\underline{Y}$ for which $\underline{X}$ is almost surely majorized by $\underline{Y}$. However, this will not be transparent from the definition ((5.12) below). We will focus attention on one particular form of stochastic majorization, the one proposed by Nevius, Proschan and Sethuraman (1977). We will mention in passing other possible definitions and refer the interested reader to the rather complicated diagram on page 316 of Marshall and Olkin (1979), which summarizes known facts about the interrelationships between the various brands of stochastic majorization.

To motivate our definition, recall that by the definition of Schur convexity (definition 2.2) we know that $\underline{x} \leq_M \underline{y}$ if and only if $g(\underline{x}) \leq g(\underline{y})$ for every Schur convex function g. Now, if $\underline{X}$ and $\underline{Y}$ are random variables, then we need to decide in what sense will $g(\underline{X})$, a random variable, provided g is Borel measurable, be required to be less than the random variable $g(\underline{Y})$. The comparison used by Nevius, Proschan and Sethuraman was based on stochastic ordering (which we will study more thoroughly in Chapter 6). The relevant concepts are defined by

Definition 5.11: Let X and Y be one dimensional random variables. We will write $X \leq_{st} Y$ (X is not stochastically larger than Y), if

$$P(X \leq x) \geq P(Y \leq x) \ \forall\ x \in \mathbb{R} .$$

Definition 5.12: Let $\underline{X}$ and $\underline{Y}$ be n-dimensional random variables. We will say that $\underline{X}$ is stochastically majorized by $\underline{Y}$ and write $\underline{X} \leq_{st.M} \underline{Y}$, if $g(\underline{X}) \leq_{st} g(\underline{Y})$ for every Borel measurable Schur convex function $g: \mathbb{R}^n \to \mathbb{R}$.

It will be recalled from Chapter 2 that in the definition of non-stochastic majorization one does not have to check every Schur convex g, in order to determine whether majorization obtains. Certain subclases suffice. For example, we could consider g's which are continuous, symmetric and convex or, instead, g's that are separable convex or in the extreme, by the HLP theorem (Theorem 2.10), g's which are separable angle functions. Now in the definition of stochastic majorization (definition 5.12), if we replace "for all Borel measurable Schur convex functions g" by for all g in some specified subclass of well behaved Schur convex functions such as those just mentioned, we find that different partial orders may be defined. Some putative definitions for stochastic majorization derived in this manner may be described as follows. First, define

$$G_1 = \{g: \ g \text{ is Borel measurable and Schur convex on } \mathbb{R}^n\} ,$$

$$G_2 = \{g: \ g \text{ is continuous, symmetric and convex on } \mathbb{R}^n\} ,$$

$$G_3 = \{g: \ g(\underline{x}) = \sum_{i=1}^{n} h(x_i), \text{ for some continuous convex function } h \text{ on } \mathbb{R}\}$$

and

$$G_4 = \{g: \ g(\underline{x}) = \sum_{i=1}^{n} (x_i - c)^+ \text{ for some } c \in \mathbb{R} \text{ or } g(\underline{x}) = \sum_{i=1}^{n} x_i\} .$$

Now, for i=1,2,3,4 we define the partial order $\le_{st.M(i)}$ by $\underline{X} \le_{st.M(i)} \underline{Y}$ iff $g(\underline{X}) \le_{st} g(\underline{Y}) \ \forall \ g \in G_i$. Of course, $\le_{st.M(1)}$ is just the same as stochastic majorization as described in definition 5.12. Since $G_4 \subset G_3 \subset G_2 \subset G_1$, it is obvious that

$$\le_{st.M(1)} \ \Rightarrow \ \le_{st.M(2)} \ \Rightarrow \ \le_{st.M(3)} \ \Rightarrow \ \le_{st.M(4)} \ .$$

Marshall and Olkin provide an example to show that $\le_{st.M(2)} \not\Rightarrow \le_{st.M(1)}$. It is not known whether $\le_{st.M(2)}$ and $\le_{st.M(3)}$ are equivalent. The relation between $\le_{st.M(3)}$ and $\le_{st.M(4)}$ is considered in exercise 11.

Faced with such a surfeit of possible definitions (Marshall and Olkin even consider 3 more choices for g), we will stick with the original choice exemplified by definition 5.12. Two equivalent versions of that definition can be obtained from the following

<u>Theorem 5.13</u>: The following conditions are equivalent:

(1) $\underline{X} \le_{st.M} \underline{Y}$

(2) $E[g(\underline{X})] \leq E[g(\underline{Y})]$ for every Schur convex g for which both expectations exist.

(3) $P(\underline{X} \varepsilon A) \leq P(\underline{Y} \varepsilon A)$ for every measurable Schur convex set A (i.e. for every A such that $\underline{x} \varepsilon A$ and $\underline{x} \leq_M \underline{y} \Rightarrow \underline{y} \varepsilon A$).

Proof: (1) => (2), since $U \leq_{st} V$ implies $E(U) \leq E(V)$ when expectations exist. (2) => (3), since the indicator function of a Schur convex set is Schur convex. (3) => (1), since one may let A = $\{\underline{x}: g(\underline{x}) > c\}$.

Stochastic majorization is preserved under mixing, normalization, a strong mode of convergence and, to a certain extent, convolution. Specifically, we have:

Theorem 5.14: Let $\{\underline{X}_\lambda: \lambda \varepsilon \Lambda\}$ and $\{\underline{Y}_\lambda: \lambda \varepsilon \Lambda\}$ be two indexed collections of random variables where Λ is a subset of $\mathbb{R}$. Let G be the distribution function of a random variable whose range is in Λ, and let $\underline{X}_G$ and $\underline{Y}_G$ be the corresponding G-mixtures of $\{\underline{X}_\lambda\}$ and $\{\underline{Y}_\lambda\}$ respectively, i.e.,

$$F_{\underline{X}_G}(\underline{x}) = \int_\Lambda P(\underline{X}_\lambda \leq \underline{x})\, dG(\lambda) \tag{5.7}$$

and $F_{\underline{Y}_G}$ is analogously defined. If $\underline{X}_\lambda \leq_{st.M} \underline{Y}_\lambda$ for every $\lambda \varepsilon \Lambda$, then $\underline{X}_G \leq_{st.M} \underline{Y}_G$.

Proof: For any function g for which $E[g(\underline{X}_G)]$ exists, it follows from (5.7) that

$$E[g(\underline{X}_G)] = \int_\Lambda E(g(\underline{X}_\lambda))\, dG(\lambda) , \tag{5.8}$$

and that an analogous expression is available for $E[g(\underline{Y}_G)]$. If g is Schur convex, then since $\underline{X}_\lambda \leq_{st.M} \underline{Y}_\lambda$, we know by condition (2) of Theorem 5.13 that $E[g(\underline{X}_\lambda)] \leq E[g(\underline{Y}_\lambda)]$, for every $\lambda \varepsilon \Lambda$. Integrating this with respect to the distribution G using equation (5.7), we conclude that $E[g(\underline{X}_G)] \leq E[g(\underline{Y}_G)]$. Since this is true for any Schur convex g for which the expectations exist, we conclude that $\underline{X}_G \leq_{st.M} \underline{Y}_G$ (again applying condition (2) of Theorem 5.13).

<u>Theorem 5.15</u>: Suppose that $\underline{X} \leq_{st.M} \underline{Y}$ and $\underline{X}' = f(\sum_{i=1}^{n} X_i)\underline{X}$, $\underline{Y}' = f(\sum_{i=1}^{n} Y_i)\underline{Y}$ where f is a Borel measurable function on $\mathbb{R}$. It follows that $\underline{X}' \leq_{st.M} \underline{Y}'$.

<u>Proof</u>: Since $\underline{X} \leq_{st.M} \underline{Y}$, it follows that $\sum_{i=1}^{n} X_i \overset{d}{=} \sum_{i=1}^{n} Y_i$ (exercise 12). Now since $\underline{X} \leq_{st.M} \underline{Y} \Rightarrow c\underline{X} \leq_{st.M} c\underline{Y}$ for any c, the theorem follows from our mixture theorem, since we have

$$F_{\underline{X}'}(\underline{x}) = \int_{-\infty}^{\infty} P(c\underline{X} \leq \underline{x})\, dG(c)$$

and

$$F_{\underline{Y}'}(\underline{x}) = \int_{-\infty}^{\infty} P(c\underline{Y} \leq x)\, dG(c)$$

where G is the common distribution of $f(\sum_{i=1}^{n} X_i)$ and $f(\sum_{i=1}^{n} Y_i)$.

<u>Theorem 5.16</u>: Suppose $\{\underline{X}_m\}$ and $\{\underline{Y}_m\}$ are two sequences of n-dimensional random variables and that $\underline{X}_m \to \underline{X}$ and $\underline{Y}_m \to \underline{Y}$ in the sense that $E(g(\underline{X}_n)) \to E(g(\underline{X}))$ for every bounded measurable function g. If $\underline{X}_m \leq_{st.M} \underline{Y}_m$ for each m, then $\underline{X} \leq_{st.M} \underline{Y}$.

<u>Proof</u>: Let g be an arbitrary Schur convex Borel measurable function. Let z be a point of continuity of both of the distributions of $g(\underline{X})$ and $g(\underline{Y})$. Observe that $I(g(\underline{x}) \leq z)$ is a bounded Borel measurable function. We may thus argue as follows.

$$\begin{aligned}
P(g(\underline{X}) \leq z) &= E(I(g(\underline{X}) \leq z)) \\
&= \lim_{m\to\infty} E(I(g(\underline{X}_m) \leq z)) \\
&= \lim_{m\to\infty} P(g(\underline{X}_m) \leq z) \\
&\geq \lim_{m\to\infty} P(g(\underline{Y}_m) \leq z) [\text{since } \underline{X}_m \leq_{st.M} \underline{Y}_m] \\
&= \lim_{m\to\infty} E(I(g(\underline{Y}_m) \leq z)) \\
&= E(I(g(\underline{Y}) \leq z)) \\
&= P(g(\underline{Y}) \leq z) .
\end{aligned}$$

Since such points z are dense in $\mathbb{R}$, it follows that $g(\underline{X}) \leq_{st} g(\underline{Y})$, and since this holds for every Borel measurable Schur convex g, we conclude $\underline{X} \leq_{st.M} \underline{Y}$.

Closure under convolution will elude us. It is possible to construct an example of random vectors $\underline{X}^{(1)}, \underline{X}^{(2)}, \underline{Y}^{(1)}$ and $\underline{Y}^{(2)}$ (even in the case n = 2) such that $\underline{X}^{(1)}, \underline{X}^{(2)}$ are independent, $\underline{Y}^{(1)}, \underline{Y}^{(2)}$ are independent, $\underline{X}^{(1)} \leq_{st.M} \underline{Y}^{(1)}$ and $\underline{X}^{(2)} <_{st.M} \underline{Y}^{(2)}$ yet $\underline{X}^{(1)} + \underline{X}^{(2)} \nleq_{st.M} \underline{Y}^{(1)} + \underline{Y}^{(2)}$. See Marshall and Olkin [1979, p. 314]. If, by chance, the random vectors always have coordinates which are in increasing order, we can get our result. Recall from (2.5) the notation

$$0_n = \{\underline{x}: x_1 \leq x_2 \leq \cdots \leq x_n\} .$$

Theorem 5.17: If $\underline{X}^{(1)}, \underline{X}^{(2)}$ are independent, $\underline{Y}^{(1)}, \underline{Y}^{(2)}$ are independ- and all four random vectors take on values restricted to the set 0_n. If $\underline{X}^{(1)} \leq_{st.M} \underline{Y}^{(1)}$ and $\underline{X}^{(2)} \leq_{st.M} \underline{Y}^{(2)}$, then $\underline{X}^{(1)} + X^{(2)} \leq_{st.M} \underline{Y}^{(1)} + \underline{Y}^{(2)}$.

Proof: If $\underline{u}, \underline{v}, \underline{w} \in 0_n$, then $\underline{u} \leq_M \underline{v} \Rightarrow \underline{u} + \underline{w} \leq_M \underline{v} + \underline{w}$. Thus, if g is Schur convex on 0_n, then for every $\underline{w} \in 0_n$, $g_{\underline{w}}(\underline{x}) = g(\underline{x} + \underline{w})$ is Schur convex on 0_n. Now, we can write for an arbitrary Schur convex g,

$$\begin{aligned} E[g(\underline{X}^{(1)} + \underline{X}^{(2)})] &= \int_{0_n} E[g(\underline{X}^{(1)} + \underline{w})]\, dF_{\underline{X}^{(2)}}(\underline{w}) \\ &\leq \int_{0_n} E[g(\underline{Y}^{(1)} + \underline{w})]\, dF_{\underline{X}^{(2)}}(\underline{w}) \\ &= E[g(\underline{Y}^{(1)} + \underline{X}^{(2)})] . \end{aligned}$$

Here we have assumed, without loss of generality, that all four random vectors $\underline{X}^{(1)}, \underline{X}^{(2)}, \underline{Y}^{(1)}$ and $\underline{Y}^{(2)}$ are independent. By conditioning on $\underline{Y}_1$, we may then prove $E[g(\underline{Y}^{(1)} + \underline{X}^{(2)})] \leq E[g(\underline{Y}^{(1)} + \underline{Y}^{(2)})]$. Since g was an arbitrary Schur convex function, we conclude $\underline{X}^{(1)} + \underline{X}^{(2)} \leq_{st.M} \underline{Y}^{(1)} + \underline{Y}^{(2)}$.

We have alluded to the fact that if $\underline{X} \leq_{st.M} \underline{Y}$, then $\sum_{i=1}^n X_i \stackrel{d}{=} \sum_{i=1}^n Y_i$. What happens if we have stochastic majorization in both directions, i.e., $\underline{X} \leq_{st.M} \underline{Y}$ and $\underline{Y} \leq_{st.M} \underline{X}$? We cannot conclude that

$\underline{X} = \underline{Y}$ or even that $\underline{X}$ and $\underline{Y}$ are identically distributed. However, it is true that they must have identically distributed order statistics. This is not surprising since an analogous result is encountered in the non-stochastic case. If $\underline{x} \leq_M \underline{y}$ and $\underline{y} \leq_M \underline{x}$ it does not follow that $\underline{x} = \underline{y}$ but it does follow that $x_{i:n} = y_{i:n}$, $i=1,2,\ldots,n$. In the stochastic case, we argue as follows.

For $j=1,2,\ldots,n$ and $z \in \mathbb{R}$ define

$$g_z^{(j)}(\underline{x}) = I(\sum_{i=1}^{j} x_{i:n} \leq z) . \tag{5.9}$$

It is clear that these $g_z^{(j)}(\cdot)$'s are Schur convex functions. Since products of Schur convex functions are again Schur convex, we conclude that for any vector $z_1, z_2, \ldots, z_n$ satisfying $z_{i+1} - z_i > z_i - z_{i-1}$, $i=1,2,\ldots,n-1$ where by convention $z_0 = 0$, we can conclude that

$$E[\prod_{j=1}^{n} g_{z_j}^{(j)}(\underline{X})] = E[\prod_{j=1}^{n} g_{z_j}^{(j)}(\underline{Y})] , \tag{5.10}$$

since both $\underline{X} \leq_{st.M} \underline{Y}$ and $\underline{Y} \leq_{st.M} \underline{X}$. However,

$$E(\prod_{j=1}^{n} g_z^{(j)}(\underline{X})) = P(\sum_{i=1}^{j} X_{i:n} \leq z_j, \ j=1,2,\ldots,n) ,$$

so that (5.10) is enough to guarantee that

$$(X_{1:n}, X_{1:n} + X_{2:n}, \ldots, X_{1:n} + X_{2:n} + \ldots + X_{n:n})$$

$$\stackrel{d}{=} (Y_{1:n}, Y_{1:n} + Y_{2:n}, \ldots, Y_{1:n} + Y_{2:n} + \ldots + Y_{n:n}) ,$$

from which it follows directly that

$$(X_{1:n}, X_{2:n}, \ldots, X_{n:n}) \stackrel{d}{=} (Y_{1:n}, Y_{2:n}, \ldots, Y_{n:n}) .$$

Note that the vector $(X_{1:n}, X_{1:n} + X_{2:n}, \ldots, X_{1:n} + X_{2:n} + \ldots + X_{n:n})$ alluded to above is just the un-normalized Lorenz curve of the vector $(X_1, \ldots, X_n)$. We can define $\leq_{st.M}$ in terms of functions of the un-normalized Lorenz curve. Thus, if we let $\mathbb{L}$ be the class of all possible un-normalized Lorenz curves, i.e.,

$$\mathbb{L} = \{(z_1, z_2, \ldots, z_n): z_{i+1} - z_i > z_i - z_{i-1}, \ i=1,2,\ldots,n\} \tag{5.11}$$

(again $z_0 = 0$ by convention), and for any random vector $\underline{X}$ we denote its un-normalized Lorenz curve by $L^*(\underline{X})$, we can verify:

Theorem 5.18: $\underline{X} \leq_{st.M} \underline{Y}$ if and only if $h(L^*(\underline{X})) \leq_{st} h(L^*(\underline{Y}))$ for every $h(\underline{z})$: $\mathbb{L} \to \mathbb{R}$ which for fixed z_n is a monotone decreasing function of each of the arguments $z_1, z_2, \ldots, z_{n-1}$.

Proof: Exercise 13.

When we introduced stochastic majorization, it was remarked that our definition of $\underline{X} \leq_{st.M} \underline{Y}$ was, in fact, equivalent to the existence of versions of $\underline{X}$ and $\underline{Y}$ for which almost sure majorization obtains. This observation, attributed by Marshall and Olkin to T. Snijders, is relatively easy to prove, if we are willing to consider rather complicated probability spaces.

Theorem 5.19: $\underline{X} \leq_{st.M} \underline{Y}$ if and only if there exist two random variables $\underline{X}', \underline{Y}'$ defined on the same probability space for which $\underline{X} \stackrel{d}{=} \underline{X}'$, $\underline{Y} \stackrel{d}{=} \underline{Y}'$ and $P(\underline{X}' \leq_M \underline{Y}') = 1$.

Proof: Let $\omega = \{(\underline{x}, \underline{y}): \underline{x} \in \mathbb{R}^n, \underline{y} \in \mathbb{R}^n, \underline{x} \leq_M \underline{y}\} \subset \mathbb{R}^{2n}$. Let μ and ν be respectively probability measures on $\mathbb{R}^n$ determined by the distributions of $\underline{X}$ and $\underline{Y}$. Since ω is a closed subset of $\mathbb{R}^{2n}$, and since Strassen's (1965) condition (30) is satisfied, there exists a probability measure λ on $\mathbb{R}^{2n}$ with marginals μ and ν which is supported by ω, i.e., with $\lambda(\omega) = 1$. Let $(\underline{X}', \underline{Y}')$ be a 2n-dimensional random variable with distribution λ, then the conditions of the theorem are clearly satisfied.

In the remainder of this chapter we will give several examples of stochastic majorization. The first is an example in which stochastic majorization occurs in the almost sure sense.

Theorem 5.20: Let $X_1, X_2, \ldots, X_n$ be positive random variables and let g be a positive star shaped function defined on $[0,\infty)$. Define $Z_i = g(X_i)$, $i=1,2,\ldots,n$. It follows that

$$P\left(\left(\sum_{i=1}^{n} X_i\right)^{-1} \underline{X} \leq_M \left(\sum_{i=1}^{n} Z_i\right)^{-1} \underline{Z}\right) = 1 . \qquad (5.12)$$

Proof: For each point w in the probability space consider a discrete uniform random variable $\hat{X}_w$ with n possible values $X_1(w), X_2(w), \ldots, X_n(w)$. Apply Theorem 4.2 (note: g star shaped implies $g(x)/x$ is non-decreasing and g is non-decreasing) and conclude that for each w

$$\left(\sum_{i=1}^{n} X_i(w)\right)^{-1} \underline{X}(w) \leq_M \left(\sum_{i=1}^{n} Z_i(w)\right)^{-1} \underline{Z}(w),$$

but this implies (5.12).

An important source of examples of stochastic majorization depends on the following preservation theorem of Nevius, Proschan and Sethuraman.

Theorem 5.21: Let $\{f(x,\lambda)\}_{\lambda>0}$ be a family of densities on $(0,\infty)$. Suppose that $f(x,\lambda)$ is totally positive of order two, i.e., for $x_1 < x_2$ and $\lambda_1 < \lambda_2$

$$\begin{vmatrix} f(x_1,\lambda_1) & f(x_1,\lambda_2) \\ f(x_2,\lambda_1) & f(x_2,\lambda_2) \end{vmatrix} \geq 0.$$

In addition, assume that $f(x,\lambda)$ satisfies the following "semi-group" property in λ

$$f(x,\lambda_1 + \lambda_2) = \int_0^\infty f(x - y,\lambda_1)\, f(y,\lambda_2)\, d\mu(y) ,$$

where μ denotes either Lebesgue measure on $(0,\infty)$ or counting measure on $\{0,1,2,\ldots\}$.

Let $(X_{\lambda_1}, X_{\lambda_2}, \ldots, X_{\lambda_n})$ denote a vector with independent components such that X_{λ_i} has density $f(x,\lambda_i)$, $i=1,2,\ldots,n$. If $\underline{\lambda} \leq_M \underline{\lambda}'$, then $X_{\underline{\lambda}} \leq_{st.M} X_{\underline{\lambda}'}$.

Proof: Without loss of generality, we may assume $n = 2$. The result will then follow, provided we can show that for any Schur convex g the function ϕ defined by

$$\phi(\lambda_1,\lambda_2) = \int_0^\infty \int_0^\infty g(x_1,x_2)\, f(x_1,\lambda_1)\, f(x_2,\lambda_2)\, d\mu(x_1)\, d\mu(x_2)$$

is again Schur convex. Suppose $\lambda_1 < \lambda_2$ and ε is small, then $(\lambda_1 + \varepsilon, \lambda_2 - \varepsilon) \leq_M (\lambda_1,\lambda_2)$, and we may write

$$\phi(\lambda_1,\lambda_2) - \phi(\lambda_1 + \varepsilon,\lambda_2 - \varepsilon)$$
$$= \int_0^\infty \int_0^\infty [f(x_1,\lambda_1)\ f(x_2,\lambda_2)$$
$$-f(x_1,\lambda_1+\varepsilon)\ f(x_2,\lambda_2-\varepsilon)]\ g(x_1,x_2)\ d\mu(x_1)\ d\mu(x_2)$$
$$= \int_0^\infty f(y,\varepsilon) \int_0^\infty \int_0^\infty f(x_1,\lambda_1)\ f(x_2 - y,\lambda_2 - \varepsilon)$$
$$-f(x_1 - y,\lambda_1)\ f(x_2,\lambda_2 - \varepsilon)]\ g(x_1,x_2)\ d\mu(x_1)d\mu(x_2)d\mu(y)\ .$$

Now, change variables and recall g is symmetric. Thus,

$$\phi(\lambda_1,\lambda_2) - \phi(\lambda_1 + \varepsilon,\lambda_2 - \varepsilon)$$
$$= \int_0^\infty f(y,\varepsilon) \iint_{z_1 \le z_2} [f(z_2,\lambda_1)f(z_1,\lambda_2-\varepsilon) - f(z_1,\lambda_1)f(z_2,\lambda_2-\varepsilon)]$$
$$\times [g(z_1+y,z_2) - g(z_1,z_2+y)]\ d\mu(z_1)\ d\mu(z_2)\ d\mu(y).$$

The Schur convexity of g guarantees that $g(z_1+y,z_2) - g(z_1,z_2+y) \le 0$, and the total positivity of order 2 of f guarantees that $f(z_2,\lambda_1)$ $f(z_1,\lambda_2-\varepsilon) - f(z_1,\lambda_1)\ f(z_2,\lambda_2-\varepsilon) \le 0$. The Schur convexity of ϕ is thus confirmed.

The hypotheses of Theorem 5.21 are strong. Four examples are available: binomial, Poisson, negative binomial and gamma. More examples can be generated by considering mixtures as follows.

Theorem 5.22: Let $X_{\underline{\lambda}}$ be a collection of n-dimensional random variables index by $\underline{\lambda} \in \mathbb{R}_n$ such that $\underline{\lambda} \le_M \underline{\lambda}' \Rightarrow \underline{X}_{\underline{\lambda}} \le_{st.M} \underline{X}_{\underline{\lambda}'}$. Let $\underline{Z}$ and $\underline{Z}'$ be two n-dimensional random vectors with corresponding distribution functions $F_{\underline{Z}}(\underline{z})$ and $F_{\underline{Z}'}(\underline{z})$. Define $\underline{U}$ and $\underline{U}'$ by

$$F_{\underline{U}}(\underline{u}) = \int_{\mathbb{R}_n} P(\underline{X}_{\underline{\lambda}} \le \underline{u})\ dF_{\underline{Z}}(\underline{\lambda})$$

and

$$F_{\underline{U}'}(\underline{u}) = \int_{\mathbb{R}_n} P(\underline{X}_{\underline{\lambda}} \le \underline{u})\ dF_{\underline{Z}'}(\underline{\lambda}).$$

If $\underline{Z} \le_{st.M} \underline{Z}'$, then $\underline{U} \le_{st.M} \underline{U}'$ [more briefly we could write $\underline{Z} \le_{st.M} \underline{Z}' \Rightarrow \underline{X}_{\underline{Z}} \le_{st.M} \underline{X}_{\underline{Z}'}$].

Proof: By Theorem 5.19 we can assume that $\underline{Z} \le_M \underline{Z}'$ a.s. Then, for any bounded Schur convex g we have $E(g(\underline{U})|\underline{Z}) \le E(g(\underline{U}')|\underline{Z}')$ which may be integrated to yield $E[g(\underline{U})] \le E[g(\underline{U}')]$. It follows that $\underline{U} \le_{st.m} \underline{U}'$.

<u>Example 5.23</u>: Let $\underline{X}_\lambda \sim$ multinomial $(N,\underline{\lambda})$ where $\lambda_i > 0$, $\sum_{i=1}^n \lambda_i = 1$, i.e.,

$$P(\underline{X}_\lambda = \underline{k}) = N! \prod_{i=1}^n \left(\frac{\lambda_i^{k_i}}{k_i!}\right),$$

$0 \le k_i \le N$, $\sum_{i=1}^n k_i = N$. If $\underline{\lambda} \le_M \underline{\lambda}'$, then $\underline{X}_\lambda \le_{st.M} \underline{X}_{\lambda'}$.
The simplest way to verify this assertion is to observe that if $Z_{\lambda_i},\ldots,Z_{\lambda_n}$ are independent random variables with $Z_i \sim$ Poisson (λ_i), then $\underline{\lambda} \le_M \underline{\lambda}'$ implies $\underline{Z}_\lambda \le_{st.M} \underline{Z}_{\lambda'}$ by Theorem 5.21. But then if we define $X_{\lambda_i} = I(\sum_{i=1}^n Z_{\lambda_i} = N)\, Z_{\lambda_i}$, we have $\underline{X}_\lambda \le_{st.M} \underline{X}_{\lambda'}$ by Theorem 5.15. But the conditional distribution of such an $\underline{X}_\lambda$ given $\underline{X}_\lambda \ne \underline{0}$ is just multinomial $(N,\underline{\lambda})$. The desired result follows, since $\sum_{i=1}^n Z_{\lambda_i} \stackrel{d}{=} \sum_{i=1}^n Z_{\lambda_i'}$ (cf. exercise 12).

Other examples are described in exercise 17.

Exercises

1. By considering the two dimensional case (m=2), verify that $X \leq_{MM} Y > X \leq_{UM} Y$.
2. Prove that $X \leq_{UM} Y \Rightarrow X \leq_{CM} Y$.
3. Prove that $X \leq_{MO} Y \Rightarrow X \leq_{MM} Y$ and that $X \leq_{CM} Y \Rightarrow X <_{MM} Y$.
4. Suppose $\underline{X}$ and $\underline{Y} \in \mathscr{L}^{(m)}$ have independent coordinate random variables such that $X_i \leq_L Y_i$, $i=1,2,\ldots,n$. Prove that $\underline{X} \leq_L \underline{Y}$.
5. Suppose $\underline{X} \in \mathscr{L}^{(m)}$ and $U \geq 0$ are independent. Assume $E(U) \in (0,\infty)$ and define $\underline{Y} = U\underline{X}$. Prove $\underline{X} \leq_L \underline{Y}$. This generalizes the misreported income model of Chapter 4.
6. Discuss the relationships between the three orders $\leq_L$, $\leq_{MOL}$ and $\leq_{gL}$ defined on $\mathscr{L}^{(m)}$.
7. Suppose that X_1 and $X_2 \in \mathscr{L}$ with corresponding Lorenz curves $L_1(u)$ and $L_2(u)$. Show that if X_1 and X_2 are independent, the Lorenz surface of the random variable (X_1,X_2) defined by (5.4) is of the form $L(u,v) = L_1(u)L_2(v)$.
8. Suppose
$$\begin{aligned} f_{X_1,X_2}(x_1.x_2) &= x_1+x_2,\ 0 < x_1 < 1,\ 0 < x_2 < 1 \\ &= 0\ , \quad \text{otherwise} \end{aligned}$$
and
$$\begin{aligned} f_{Y_1,Y_2}(y_1,y_2) &= 1,\ 0 < x_1 < 1,\ 0 < x_2 < 1 \\ &= 0\ , \quad \text{otherwise}\ . \end{aligned}$$
Evaluate and compare the corresponding Lorenz surfaces defined by (5.4). What happens if we use the Taguchi definition (5.3)?
9. In accordance with custom, inequality measures vary from 0 to 1 with 1 representing the unachievable case of greatest inequality. In equation (5.6) what value should be taken for c_m in order to have a measure satisfying $0 \leq G^{(m)}(\underline{X}) < 1$?
10. Verify that the matrix P given in equation (5.2) is not expressible as a finite product of Robin Hood matrices. [Verify

that finite products of Robin Hood matrices, in the 3×3 case, cannot have 3 zero elements.]

11. Determine whether $\leq_{st.M(4)} \Rightarrow \leq_{st.M(3)}$.

12. Suppose $\underline{X} \leq_{st.M} \underline{Y}$. Prove that $\sum_{i=1}^{n} X_i \overset{d}{=} \sum_{i-1}^{n} Y_i$.

13. Prove Theorem 5.18.

14. Prove that the random variables $\underline{X}'$ and $\underline{Y}'$ alluded to in Theorem 5.19 cannot be independent.

15. Suppose $\underline{X} \leq_{st.M} \underline{Y}$. Prove that for any $\underline{a}$ with $\sum_{i=1}^{n} a_i = 1$,

$$P\Big(\Big(\sum_{i=1}^{n} X_i\Big)^{-1} \underline{X} \geq_M \underline{a}\Big) \leq P\Big(\Big(\sum_{i=1}^{n} Y_i\Big)^{-1} \underline{Y} \geq_M \underline{a}\Big).$$

16. Suppose $X_1,\ldots,X_n$ are independent random variables with X_i Poisson (λ_i) and that $X_1^*,\ldots,X_n^*$ are i.i.d. Poisson $(\bar{\lambda})$ random variables $\big(\bar{\lambda} = \frac{1}{n} \sum_{i=1}^{n} \lambda_i\big)$. Define $g(\underline{x}) = \big(\frac{1}{n} \sum_{i=1}^{n} x_i^2\big)/\big(\frac{1}{n} \sum_{i=1}^{n} x_i\big)^2$. Verify that $P(g(\underline{X}) > \delta) \leq P(g(\underline{X}^*) > \delta)$, $\delta > 0$.

17. a) Show that the family of n-dimensional Dirichlet random variables $\{\underline{X}_{\underline{\lambda}}\}$ satisfies $\underline{\lambda} \leq_M \underline{\lambda}' \Rightarrow \underline{X}_{\underline{\lambda}} \leq_{st.M} \underline{X}_{\underline{\lambda}'}$.

b) Multivariate negative binomial. Consider a sequence of independent experiments each with (n+1) possible outcomes 0,1,2,...,n with associated probabilities

$$\Big(\frac{\lambda_0}{\sum_{i=0}^{n} \lambda_i}, \frac{\lambda_1}{\sum_{i=0}^{n} \lambda_i}, \ldots, \frac{\lambda_n}{\sum_{i=0}^{n} \lambda_i}\Big).$$

Define $\underline{X}_{\underline{\lambda}} = (X_1,\ldots,X_n)$ by X_i = the number of outcomes of type i that precede the N'th outcome of type 0 (where N is a fixed integer). Let $\underline{\lambda} = (\lambda_1,\ldots,\lambda_n)$. Prove that $\underline{\lambda} \leq_M \underline{\lambda}' \Rightarrow$ $\underline{X}_{\underline{\lambda}} \leq_{st.M} \underline{X}_{\underline{\lambda}'}$.

CHAPTER 6

SOME RELATED ORDERINGS

In this chapter we consider certain partial orders defined on $\mathscr{L}$ (the class of non-negative random variables with positive finite expections) that are closely related to the Lorenz ordering (Chapter 3). The first, *-ordering, is often easier to deal with than Lorenz ordering, and can sometimes be used to verify Lorenz ordering which it implies. The other group of partial orderings to be discussed are those known as stochastic dominance of degree k, k=1,2,.... Degree 1 is just stochastic ordering. Degree 2 is intimately related to the Lorenz order, but distinct. Higher degree stochastic orders are most frequently encountered in economic contexts. The treatment provided here is brief.

6.1 Star-ordering

Let $X,Y \in \mathscr{L}$ with corresponding distribution functions F_X and F_Y. Star-shaped ordering or more briefly, star ordering is defined as follows.

Definition 6.1: We say that X is star-shaped with respect to Y, and write $X \leq_* Y$ if $F_X^{-1}(u)/F_Y^{-1}(u)$ is a non-increasing function of u.

Definition 6.1 is slightly more general than the definition provided by Marshall and Olkin. In order to have $X \leq_* Y$, they require that "$F_Y^{-1}(F_X(x))/x$ be increasing". This implies but is not implied by "$F_X^{-1}(u)/F_Y^{-1}(u)$ is non-increasing". For example, it is possible to have $X \leq_* Y$ for certain pairs of discrete random variables using Definition 6.1, whereas discrete random variables cannot be *-ordered using the Marshall-Olkin definition.

Since $F_{cX}^{-1}(u) = cF_X^{-1}(u)$ for any positive c and any $X \in \mathscr{L}$, it is obvious that *-ordering is scale invariant. A simple sufficient condition for $F_Y^{-1}(F_X(x))/x$ to be increasing is that $F_Y^{-1}(F_X(x))$ be convex on the support of F_X (see exercise 1).

If $F_X^{-1}(u)/F_Y^{-1}(u)$ has a simple differentiable form, then we can check for *-ordering by verifying that its derivative is non-positive. We can use *-ordering to verify that Lorenz ordering obtains as a consequence of the following theorem.

<u>Theorem 6.2</u>: Suppose $X,Y \in \mathscr{L}$. If $X \leq_* Y$, then $X \leq_L Y$.

<u>Proof</u>: Without loss of generality, since both partial orders are scale invariant, we may assume that $E(X) = E(Y) = 1$. In such a case

$$L_X(u) - L_Y(u) = \int_0^u [F_X^{-1}(v) - F_Y^{-1}(v)]\, dv \,. \tag{6.1}$$

Since $F_X^{-1}(v)/F_Y^{-1}(v)$ is non-increasing, the integrand is first positive and then negative as v ranges from 0 to 1 [cf. the proof of Theorem 4.1]. Thus, the integral assumes its smallest value when $u = 1$. It follows that $L_X(u) - L_Y(u) \geq L_X(1) - L_Y(1) = 1 - 1 = 0$, and consequently $X \leq_L Y$.

If we return now to the proof of Theorem 4.1, it is clear that conditions (ii) of that theorem actually are sufficient for $g(X) \leq_* X$ $\forall\, X \in \mathscr{L}$, and the theorem could be restated as follows

<u>Theorem 6.3</u>: Let $g: \mathbb{R}^+ \to \mathbb{R}^+$. The following are equivalent.

(i) $g(X) \leq_L X, \quad \forall\, X \in \mathscr{L}$

(ii) $g(X) \leq_* X, \quad \forall\, X \in \mathscr{L}$

(iii) $g(x) > 0\ \forall\, x > 0$, $g(x)$ is non-decreasing on $[0,\infty)$ and $g(x)/x$ is non-increasing on $(0,\infty)$.

Theorem 4.2 can be analogously restated. If we return again to the proof of Theorem 6.2, we can see that the important consequence of *-ordering was that $F_X^{-1}(v) - F_Y^{-1}(v)$ had only one sign change (+,-) on the interval [0,1]. This sign-change property is not scale invariant, but it does permit us to formulate the following simple sufficient condition for Lorenz ordering.

Theorem 6.4: Suppose $X,Y \in \mathscr{L}$ and that $[F_X^{-1}(v)/E(X)] - [F_Y^{-1}(v)/E(Y)]$ has at most one sign change (from + to -) as v ranges from 0 to 1. It follows that $X \leq_L Y$.

Proof: The result is obvious from the discussion following (6.1).

If inverse functions cross at most once then the same holds for the original functions. Consequently, we can restate the last sufficient condition in the form: A sufficient condition that $X \leq_L Y$ is that $F_X(E(X)x) - F_Y(E(Y)x)$ has at most one sign change (from - to +) as x ranges from 0 to ∞. If $X \leq_* Y$ one can show that $F_X(\lambda x) - F_Y(\nu x)$ has at most one sign change (from - to +) for any choices of λ and ν. It is thus evident that if we use the sign-change property alluded to in Theorem 6.4 to define a partial order on $\mathscr{L}$, it will occupy an intermediate position between $\leq_*$ and $\leq_L$.

Definition: We will say that X is sign-change ordered with respect to Y and write $X \leq_{s.c.} Y$, if $[F_X^{-1}(v)/E(X)] - [F_Y^{-1}(v)/E(Y)]$ has at most one sign change (from + to -) as v ranges from 0 to 1.

A simple sufficient condition for sign change ordering can be stated in terms of density crossings (assuming densities exist). Thus,

Theorem 6.5: Let $X,Y \in \mathscr{L}$ have corresponding densities $f_X(x)$ and $f_Y(x)$ (with respect to a convenient dominating measure on $\mathbb{R}^+$, in the most abstract setting). If the function

$$E(X)f_X(E(X)x) - E(Y)f_Y(E(Y)x) \tag{6.2}$$

has two sign changes (from - to + to -) as x ranges from 0 to ∞, then $X \leq_{s.c.} Y$.

Proof: (6.2) is merely the density of $X/E(X)$ minus the density of $Y/E(Y)$. The difference between the distribution functions

$$F_X(E(X)x) - F_Y(E(Y)x) = F_{X/E(X)}(x) - F_{Y/E(Y)}(x)$$

has the sign sequence $-,+$ since it is a function which begins at 0 when $x = 0$, ends at 0 (as $x \to \infty$) and whose derivative by (6.2) has sign sequence $-,+,-$.

Verification that $X \leq_L Y$ is frequently most easily done by using the density crossing argument (Theorem 6.5) or by showing *-ordering obtains (if $F_X^{-1}(u)$ and $F_Y^{-1}(u)$ are available in convenient tractable forms).

Example 6.6: If $X \leq_* Y$, then $X_{i:n} \leq_* Y_{i:n}$ for every $i \leq n$. This follows since $X_{i:n} \overset{d}{=} F_X^{-1}(U_{i:n})$ where $U_{i:n}$ is the i'th order statistic from a sample of size n from a uniform (0,1) distribution. Similarly, $Y_{i:n} \overset{d}{=} F_Y^{-1}(U_{i:n})$. So if we let $G_{i:n}$ be the distribution function of $U_{i:n}$, we can write

$$\frac{F_{X_{i:n}}^{-1}(u)}{F_{Y_{i:n}}^{-1}(u)} = \frac{F_X^{-1}[G_{i:n}^{-1}(u)]}{F_Y^{-1}[G_{i:n}^{-1}(u)]} .$$

But this ratio is non-increasing as u increase, since $G_{i:n}^{-1}$ is monotone increasing and $F_X^{-1}(v)/F_Y^{-1}(v)$ is a non-increasing function of v (since $X \leq_* Y$).

Example 6.7: Suppose X has a symmetric distribution on the interval $[0,c]$. The sample median (for samples of odd size) decreases in inequality in the Lorenz sense as sample size increases. Specifically, if $m < m'$

$$X_{m'+1:2m'+1} \leq_L X_{m+1:2m+1} .$$

If we consider $m' = m+1$, and denote the density of $X_{m+1:2m+1}$ by $f_m(x)$, we find that the ratio of densities $f_{m+1}(x)/f_m(x)$ is given by

$$\frac{f_{m+1}(x)}{f_m(x)} = F_X(x)[1 - F_X(x)] \frac{2(2m+3)}{(m+1)} . \tag{6.3}$$

By symmetry $E(X_{m+1:2m+1}) = E(X_{m+2:2m+3}) = c/2$, so that we do not have to corect for differences in means. Since expression (6.3) is greater than 1 for intermediate values of x and less than 1 for large and small values, it follows that, by Theorem 6.5, $X_{m+2:2m+3} \leq_{s.c.} X_{m+1:2m+1}$, and consequently they are similarly Lorenz ordered.

6.2 Stochastic dominance

The initial work on stochastic dominance was executed in the context of bounded non-negative random variables. Subsequent researchers (especially Fishburn (1980)) have extended consideration to unbounded random variables but subtle changes are required in the definitions and interpretations. In the present development we will content ourselves with consideration of the bounded case. It may be and frequently has been remarked that restriction to bounded random variables is not a serious restriction when dealing with real world data.

For the remainder of this section all random variables are non-negative and are assumed to be bounded above by $B > 0$. We will denote the class of such random variables by $\mathscr{L}_B$. Evidently if $X \in \mathscr{L}_B$, $E(X^k) < \infty \;\forall\; k$. Associated with $X \in \mathscr{L}_B$ is a sequence of "distribution" functions defined by repeated integration. Thus for $x \in [0,B]$,

$$\tilde{F}_X^{(1)}(x) = P(X \leq x) = F_X(x) \tag{6.4}$$

and for $i=2,3,\ldots$

$$\tilde{F}_X^{(i)}(x) = \int_0^x \tilde{F}_X^{(i-1)}(y)\, dy. \tag{6.5}$$

The word "distribution" is placed in quotation marks because, although $F_X^{(i)}(x)$ $(i=2,3,\ldots)$ is continuous and non-decreasing, it has not been normalized to be a true distribution function, i.e. in general we do not have $\tilde{F}_X^{(i)}(B) = 1$. An interpretation of the i'th distribution" function is available as follows.

$$\tilde{F}_X^{(i)}(x) = \int_0^x \tilde{F}_X^{(i-1)}(y_{i-1})\, dy_{i-1}$$

$$= \int_0^x dy_{i-1} \int_0^{y_{i-1}} F_X^{(i-2)}(y_{i-2})\, dy_{i-2}$$

$$= \ldots$$

$$= \int_0^x dy_{i-1} \int_0^{y_{i-1}} dy_{i-2} \ldots \int_0^{y_2} dy_1 \int_0^{y_1} dF_X(y_0).$$

The range of integration is

$$\{(y_0,\ldots,y_{i-1}):\ y_0 \le y_1 \le y_2 \ldots \le y_{i-1} \le x\}.$$

Performing the integration in reverse order we have:

$$\tilde{F}_X^{(i)}(x) = \int_0^x dF_X(y_0) \int_{y_0}^x dy_1 \int_{y_1}^x dy_2 \ldots \int_{y_{i-2}}^x dy_{i-1}$$

$$= \frac{1}{(i-1)!} \int_0^x (x-y_0)^{i-1}\, dF_X(y_0). \tag{6.6}$$

It is thus possible to write $\tilde{F}_X^{(i)}(x)$ as a function of the moment distributions of F_X. The j'th moment distribution of X is defined by

$$F_X^{(j)}(x) = \int_0^x x^j\, dF_X(x)/E(X^j). \tag{6.7}$$

($F_X^{(1)}$ was introduced in Chapter 3 in the discussion of Lorenz curves). From (6.6) we have

$$\tilde{F}_X^{(i)}(x) = \frac{1}{(i-1)!} \sum_{j=0}^{i-1} \binom{i-1}{j} x^{i-1-j}(-1)^j E(X^j) F_X^{(j)}(x). \tag{6.8}$$

Observe that setting x = B in (6.8) yields,

$$\tilde{F}_X^{(i)}(B) = \frac{1}{(i-1)!} E[(B-X)^{i-1}]. \tag{6.9}$$

Stochastic dominance is defined in terms of the sequence of "distribution" functions (6.5).

<u>Definition</u>: Let X and Y ε $\mathscr{L}_B$. For n = 1,2,... we say that X is n'th degree stochastically dominated by Y and write $X \le_{s.d.(n)} Y$, if

$$\tilde{F}_X^{(j)}(B) \ge \tilde{F}_Y^{(j)}(B),\ j = 1,2,\ldots n-1 \tag{6.10}$$

and

$$\tilde{F}_X^{(n)}(x) \geq \tilde{F}_Y^{(n)}(x), \; \forall x \in [0,B]. \tag{6.11}$$

Fishburn's (1976, 1980) variant definition of n'th degree stochastic dominance involves only condition (6.11) (see exercise 7). The present definition is in the spirit of Whitmore and Findlay (1978). Stochastic dominance of degrees 2 and 3 have been much used in decision making contexts. Observe that first degree stochastic dominance is just the usual stochastic ordering denoted earlier by $\leq_{st}$.

In certain economic, financial and decision theoretical contexts interest centers on expectations of the form E(u(X)) where u is a completely monotone function in the following sense.

Definition: u: $\mathbb{R}^+ \to \mathbb{R}^+$ is completely monotone, if it is infinitely differentiable and if

$$(-1)^{n+1}\frac{d^n}{dx^n}\,u(x) \geq 0 \quad \forall\; x > 0$$

for every n.

A completely monotone function is thus, succinctly, one whose successive derivatives alternate in sign according to the sequence +,-,+,... . A remarkable observation found, for example, in Feller (1971, p. 415), is that such functions are necessarily Laplace transforms. Specifically, if u is completely monotone, then there exist two distribution functions G_1 and G_2 on $[0,\infty)$ such that for some c_1 and c_2

$$u(x) = c_1 \int_0^\infty e^{-xy}\, dG_1(y) - c_2 \int_0^\infty e^{-xy}\, dG_2(y). \tag{6.12}$$

Less restrictive conditions on u would involve requirements that the signs of the successive derivatives of u alternate, but just up to a point. For n=1,2,... we may define classes $\{U_n\}_{n=1}^\infty$ of functions as follows:

$$U_n = \{u \,|\, u: \mathbb{R}^+ \to \mathbb{R}^+,\; (-1)^{k+1}D^k u \geq 0,\; k=1,2,\ldots n\} \tag{6.13}$$

where D^k is the k'th derivative operator. Thus, U_1 consists of the differentiable non-decreasing functions. U_2 consists of twice differentiable non-decreasing concave functions. U_3 consists of members of U_2 which have a non-negative third derivative, etc. Evidently,

$$U_n \subset U_{n-1}, \ n=2,3,\ldots \ . \tag{6.14}$$

Our definitions of stochastic dominance of degrees 1,2,... assume a particularly simple form when related to expectations of functions in the classes $U_1.U_2,\ldots$.

Theorem 6.8: For $X,Y \in \mathscr{L}_B$, $X \leq_{s.d.(n)} Y$ if and only if $E(u(X)) \leq E(u(Y))$ for every $u \in U_n$ for which the expectations exist.

Proof: Suppose $X \leq_{s.d.(n)} Y$ and $u \in U_n$. Repeated integration by parts yields

$$\begin{aligned} E(u(Y)) - E(u(X)) &= \int_0^B u(x)\, d[F_Y(x) - F_X(x)] \\ &= u(x)\ [\tilde{F}_Y^{(1)}(x) - \tilde{F}_X^{(1)}(x)]\Big|_0^B \\ &\quad - u'(x)\ [\tilde{F}_Y^{(2)}(x) - \tilde{F}_X^{(2)}(x)]\Big|_0^B \\ &\quad + \ldots \\ &\quad + (-1)^{n-1} u^{(n-1)}(x)\ [\tilde{F}_Y^{(n)}(x) - \tilde{F}_X^{(n)}(x)]\Big|_0^B \\ &\quad + (-1)^n \int_0^B u^{(n)}(x) [\tilde{F}_Y^{(n)}(x) - \tilde{F}_X^{(n)}(x)]dx \end{aligned}$$

where $u^{(i)}(x) = D^i u(x)$. However all these terms are non-negative by (6.10) and (6.11).

To prove the converse, we write, as above

$$\begin{aligned} E(u(Y) - u(X)) &= \sum_{i=0}^{n-1} (-1)^i\ u^{(i)}(B)\ [\tilde{F}_Y^{(i+1)}(B) - \tilde{F}_X^{(i+1)}(B)] \\ &\quad + (-1)^n \int_0^B u^{(n)}(x)\ [\tilde{F}_Y^{(n)}(x) - \tilde{F}_X^{(n)}(x)]dx. \end{aligned}$$

If (6.10) fails to hold for some j, then we will have $E(u(Y) - u(X)) < 0$ for any $u \in U_n$ for which $u^{(j)}(B)$ is much bigger than the other $u^{(i)}(B)$'s and $u^{(n)}(x)$. If (6.11) fails to hold at some point x_0, we will have $E(u(Y) - u(X)) < 0$ for a $u \in U_n$ for which $u^{(n)}(x_0)$ is large compared with other values of $u^{(n)}(x)$ and for which the $u^{(i)}(B)$'s, $i < n$ are small. It is tedious to construct such members of U_n, but it can be done. The cases n=1,2 and 3 have been frequently dealt with in the literature.

In the light of (6.9), condition (6.10) is seen to involve an inequality involving complicated functions of the first n-2 moments of X and Y. For large values of n it is difficult to interpret the implications of such a condition. The Fishburn variant definition (see exercises 3 and 7) avoids this problem. The resulting partial order, requiring (6.11) only, is interpretable in terms of a slightly smaller class of utility functions say $\tilde{U}_n$, a proper subset of U_n (again see exercise 7). The price we pay is that although utility classes $\{U_n\}$ are readily motivated in terms involving risk aversion, it is not clear why one should be happy to restrict attention to the more structured classes $\{\tilde{U}_n\}$.

An intimate relationship between second degree stochastic dominance and Lorenz ordering is immediately evident.

Theorem 6.9: Let $X, Y \in \mathscr{L}_B$. $X \leq_L Y$ if and only if $Y/E(Y) \leq_{s.d.(2)} X/E(X)$.

Proof: Without loss of generality, assume $E(X) = E(Y) = 1$. Suppose $X \leq_L Y$ and consider $u \in U_2$. Since u is concave, -u is convex and by by Theorem 3.2, $E(u(Y)) \leq E(u(X))$. Since u was arbitrary in U_2, we conclude that $Y \leq_{s.d.(2)} X$.

Suppose $Y \leq_{s.d.(2)} X$. Consider angle functions of the form

$$\begin{aligned} g_c(x) &= -x + c, \quad x \leq c \\ &= 0, \quad\quad\quad x > c. \end{aligned}$$

The functions $-g_c$ can be approximated arbitarily closely by members of U_2 and, consequently, $E[g_c(X)] \leq E(g_c(Y))$ for every c. But this is enough to ensure that $X \leq_L Y$.

A caveat is in order regarding the application of Theorem 6.9. It says that Lorenz ordering and stochastic dominance of degree 2 are intimately related. It does <u>not</u> say that one is the reverse of the other. To see this, first observe that from Theorem 6.8 we know that

$$X \leq_{s.d.(n)} Y \Rightarrow X \leq_{s.d.(n+1)} Y. \tag{6.15}$$

Also note that if $X \leq_{s.d.(n)} Y$ and $Y \leq_{s.d.(n)} X$, we must have $X \stackrel{d}{=} Y$. This already tells us that stochastic dominance of degree 2 is distinctly different from reverse Lorenz ordering, since Lorenz ordering is scale invariant, i.e. if $X \leq_L Y$ and $Y \leq_L X$ we can only conclude $X \stackrel{d}{=} cY$ for some c.

Now we seek a pair of random variables, not identically distributed, for which $X \leq_L Y$ and $X \leq_{s.d.(1)} Y$. For such random variables we will have $X \leq_L Y$ yet $Y \not\leq_{s.d.(2)} X$. A simple example is provided by considering $X \sim$ uniform (1,1.5) and $Y \sim$ uniform (2,4). Clearly $X \leq_L Y$, $X \leq_{s.d.(1)} Y$, $X \leq_{s.d.(2)} Y$ and $Y \not\leq_{s.d.(2)} X$. In this example we also have $X \leq_{s.d.(2)} Y$, yet $Y \not\leq_L X$. Thus division by expectations in Theorem 6.9 is indispensible.

What can be said about preservation and attenuation of stochastic dominance? The analog of Theorem 4.1, is not very interesting.

<u>Theorem 6.10</u>: Let $g: [0,B] \to [0,B]$. The following are equivalent

(i) $g(X) \leq_{s.d.(n)} X \ \forall X \in \mathscr{L}_B$

(ii) $g(x) \leq x, \ \forall x \in [0,B]$.

Neither is the characterization of functions which preserve stochastic dominance.

<u>Theorem 6.11</u>: Let $g: [0,B] \to [0,B]$. A necessary and sufficient condition that $X \leq_{s.d.(n)} Y$ imply $g(X) \leq_{s.d.(n)} g(Y)$ is that g is non-decreasing on [0,B].

In both cases the necessity part of the proof involves considering degenerate X's and Y's. Among degenerate random variables the stochastic dominance relations of differing degrees coincide, i.e., if $P(X=a) = P(Y=b) = 1$, then $X \leq_{s.d.(n)} Y \Leftrightarrow a \leq b$.

Theorems 6.10 and 6.11 reinforce the evidence that stochastic dominance of degree 2 is a different breed of cat than Lorenz ordering.

Exercises

1. Let F_X and F_Y be (for convenience) strictly increasing distribution functions on $(0,\infty)$. We say that X is convex with respect to Y and write $X \leq_c Y$, if $F_Y^{-1}(F_X(x))$ is a convex function on $(0,\infty)$. Prove that if $X \leq_c Y$ then necessarily $X \leq_* Y$.
2. Construct examples to verify that convex ordering, sign change ordering, star ordering and Lorenz ordering are distinct partial orders.
3. Compare the present definition of stochastic dominance of order 3 with that used in Fishburn (1976). (As in exercise 7 below with $\alpha = 3$, but restricted to $\mathscr{L}_B$).
4. Prove the converse of Theorem 6.8 when $n = 1,2$.
5. Prove that $X \leq_* Y$ if and only if for every $c > 0$, X and cY have distribution functions which cross at most once.
6. Show by example that the converse to Theorem 6.5 does not hold.
7. (Fishburn's stochastic dominance for unbounded random variables). Let X and Y be non-negative random variables. We write $X \leq_{s.d.(\alpha)} Y$, if

$$F_X^{(\alpha)}(x) \geq F_Y^{(\alpha)}(x) \quad \forall\, x \geq 0$$

where

$$F_X^{(\alpha)}(x) = \frac{1}{\Gamma(\alpha)} \int_0^x (x-y)^{\alpha-1}\, dF_X(y), \qquad \alpha \geq 1.$$

(a) Verify that for $\alpha < \alpha'$

$$F_X^{(\alpha')}(x) = \frac{1}{\Gamma(\alpha'-\alpha)} \int_0^x (x-y)^{\alpha'-\alpha-1} F_X^{(\alpha)}(y)\, dy$$

and conclude that α dominance implies α' dominance.

(b) Let $V_\alpha = \{v$: v is real valued continuous on $[0,\infty)$, positive on $(0,\infty)$ and satisfies $\int_0^\infty x^{\alpha-1} v(x)\, dx < \infty\}$.

Define $\tilde{U}_\alpha$ by: $u \in \tilde{U}_\alpha$, if there exists some $v \in V_\alpha$ and some real c such that

$$u(x) = -\int_x^\infty v(y)(y-x)^{\alpha-1}\, dy + c.$$

Show that $X \leq_{s.d.(\alpha)} Y$ if and only if $E(u(X)) \leq E(u(Y))$ for every $u \in \tilde{U}_\alpha$.

8. X and Y are said to be ordered in dispersion $(X \leq_{disp} Y)$, if $F_X^{-1}(\beta) - F_X^{-1}(\alpha) < F_Y^{-1}(\beta) - F_Y^{-1}(\alpha)$ whenever $0 < \alpha < \beta < 1$. Show that $X \leq_{disp} Y$ if and only if for every real c the distribution functions of X + c and Y cross at most once, and if there is a sign change, $F_{X+c} - F_Y$ changes sign from - to + (Shaked, 1982). Compare this result with that of exercise 5.

9. If X has a symmetric density on (0,2) and $f_X(x)$ is decreasing on (1,2), show that $var(X) \leq 1/3$. [Compare X with a uniform (0,2) random variable].

CHAPTER 7

SOME APPLICATIONS

A glance at the table of contents of Marshall and Olkin's book will indicate that approximately 40 percent of the book (i.e., 200 pages) is devoted to applications. Clearly, we will make no attempt to duplicate such an exhaustive list. We will content ourselves with ten small examples which, it is hoped, will hint at the breadth of the areas of possible application. For more, of course, the reader will need to consult Marshall and Olkin (Chapters 7-10 and 12-13).

7.1. A geometric inequality of Cesaro

Let ℓ_1, ℓ_2 and ℓ_3 denote the lengths of the three sides of a triangle. Cesaro is credited with the observation that for any triangle its side lengths satisfy

$$\ell_1\ell_2\ell_3 \leq \frac{1}{8}(\ell_1 + \ell_2)(\ell_2 + \ell_3)(\ell_3 + \ell_1) . \tag{7.1}$$

To prove this, we merely demonstrate that the left hand side and right hand side are just a particular Schur convex function evaluated at two points in $\mathbb{R}^3$ which are related by majorization. To this end, consider

$$\underset{\sim}{\tilde{\ell}} = \begin{pmatrix} 1/2 & 1/2 & 0 \\ 0 & 1/2 & 1/2 \\ 1/2 & 0 & 1/2 \end{pmatrix} \underset{\sim}{\ell} . \tag{7.2}$$

Since the matrix in (7.2) is doubly stochastic, we have $\underset{\sim}{\tilde{\ell}} \leq_M \underset{\sim}{\ell}$. Now consider the function $g(\underset{\sim}{x}) = -\prod_{i=1}^{3} x_i$. It is easy to verify that g is Schur convex (apply Schur's condition, Theorem 2.5). Inequality (7.1) then follows readily. But, in retrospect, it had nothing to do with the fact that ℓ_1, ℓ_2, ℓ_3 were lengths of sides of a triangle! If indeed the ℓ_i's were lengths of sides of a triangle, we would know that the largest of the three would have to be no larger than the sum of the other two, i.e.,

$$\ell_{3:3} \le \ell_{1:3} + \ell_{2:3}, \tag{7.3}$$

and, of course, $\ell_{1:3} \ge 0$. It follows that if the ℓ_i's are sides of a triangle, we have

$$(\ell_1, \ell_2, \ell_3) \le_M (0, \tfrac{p}{2}, \tfrac{p}{2}) \tag{7.4}$$

where $p = \ell_1 + \ell_2 + \ell_3$ is the perimeter of the triangle. Equation (7.4) does not hold for any vector (ℓ_1, ℓ_2, ℓ_3), only for vectors $\underline{\ell}$ which satisfy (7.3). If we evaluate the Schur convex function $h(\underline{x}) = -(x_1 + x_2)(x_2 + x_3)(x_3 + x_1)$ at each of the vectors in (7.4), we conclude that

$$(\ell_1 + \ell_2)(\ell_2 + \ell_3)(\ell_3 + \ell_1) \ge p^3/4 \tag{7.5}$$

for any triangle (actually the inequality is strict if we consider only non-degenerate triangles). In addition, for any vector (ℓ_1, ℓ_2, ℓ_3) including sides of a triangle we have

$$(\tfrac{p}{3}, \tfrac{p}{3}, \tfrac{p}{3}) \le_M (\ell_1, \ell_2, \ell_3) \tag{7.6}$$

where again $p = \ell_1 + \ell_2 + \ell_3$. Evaluating $h(\underline{x})$ at each of the points in (7.6) yields

$$(\ell_1 + \ell_2)(\ell_2 + \ell_3)(\ell_3 + \ell_1) \le 8p^3/27 \tag{7.7}$$

with equality in the case of an equilateral triangle.

7.2. Matrices with prescribed characteristics roots

The relationship between the diagonal elements of an $n \times n$ matrix and the vector of its n eigenvalues was early discovered to involve majorization. Schur showed that if A is an $n \times n$ Hermitian matrix, then its diagonal vector is necesarily majorized by the vector of its eigenvalues. A more recent theorem discovered independently by Horn (1954) and Mirsky (1958) is the following.

Theorem 7.1: Let $\underline{a}$ and $\underline{w}$ be two vectors in $\mathbb{R}^n$. If $a \leq_M \underline{w}$, then there exists a real symmetric $n\times n$ matrix A with $\underline{a}$ as its diagonal vector and $\underline{w}$ as its vector of eigenvalues.

To capture the basic idea of the proof without too much algebraic complexity, we will consider the two preliminary lemmas only in the case $n=3$. First we have

Lemma 7.2: If $\underline{a}, \underline{b} \in \mathbb{R}^3$ are such that $\underline{a} \leq_M \underline{b}$, then there exist numbers c_2 and c_3 such that $b_{1:3} \leq c_2 \leq b_{2:3} \leq c_3 \leq b_{3:3}$ and $(a_{2:3}, a_{3:3}) \leq_M (c_2, c_3)$.

Proof. Without loss of generality $a_1 + a_2 + a_3 = b_1 + b_2 + b_3 = 1$ and $a_1 \leq a_2 \leq a_3$, $b_1 \leq b_2 \leq b_3$. Since $\underline{a} \leq_M \underline{b}$, we have $a_1 \geq b_1$, $a_1 + a_2 \geq b_1 + b_2$ and $b_3 \geq a_3$. Obviously $b_1 \leq a_2$ and $a_3 \leq b_3$. We consider 3 cases.

Case (i). If $a_2 \leq b_2 \leq a_3$, we may take $c_2 = a_2$ and $c_3 = a_3$ and we are done.

Case (ii). Suppose $a_2 \leq a_3 < b_2$. We seek c_2, c_3 such that $c_2 + c_3 = a_2 + a_3$, $a_2 \geq c_2$ and $b_1 \leq c_2 \leq b_2 \leq c_3 \leq b_3$. Thus we wish to choose $c_2 \in (b_1, a_2)$. An acceptable choice is any c_2 in the interval $(b_1, a_1 + a_2 - b_3)$ (and $c_3 = a_2 + a_3 - c_2$).

Case (iii). Suppose $b_2 < a_2 \leq a_3$. In this case an acceptable choice is $c_2 = b_2$ and $c_3 = a_2 + a_3 - b_2$.

Lemma 7.3: If the real numbers (w_1, w_2, w_3) and (α_2, α_3) satisfy

$$w_1 \leq \alpha_2 \leq w_2 \leq \alpha_3 \leq w_3, \tag{7.8}$$

then there exists a real symmetric matrix of the form

$$A = \begin{pmatrix} \alpha_3 & 0 & p_3 \\ 0 & \alpha_2 & p_2 \\ p_3 & p_2 & p_1 \end{pmatrix} \tag{7.9}$$

whose eigenvalues are w_1, w_2 and w_3.

Proof: If we successively substitute w_1, w_2 and w_3 for λ in the determinantal equality $|A - \lambda I| = 0$, we obtain three linear equations in three unknowns p_1, p_2^2 and p_3^2. Subject to the constraint (7.8), it is not difficult to verify that a solution of the form $p_1 \in \mathbb{R}$, $p_2^2 \geq 0$, $p_3^2 \geq 0$ does exist. Hence, we can find p_1, p_2 and p_3 with the desired properties.

n-dimensional versions of Lemmas 7.2 and 7.3 were obtained by Mirsky (1958). Using them, an inductive proof of Theorem 7.1 is readily obtainable.

7.3. Variability of sample medians and means

Let $X_1, X_2, \ldots X_n$ be i.i.d. non-negative random variables with corresponding order statistics $X_{1:n}, X_{2:n}, \ldots, X_{n:n}$. If $n = 2m+1$ is odd, then the sample median is $X_{m+1:2m+1}$. As m increases, it is plausible that the sample median will tend to concentrate more and more closely near to the population median, i.e., $F_X^{-1}(\frac{1}{2})$. For convenience we will assume that the common distribution of the X_i's is absolutely continuous with corresponding probability density function $f_X(x)$. Yang (1982) showed that $\mathrm{var}(X_{m+1:2m+1})$ is less than or equal to var(X). In the case of a symmetric density, i.e., f_X of the form

$$\begin{aligned} f_X(x) &= f_X(c - x), \quad 0 < x < c , \\ &= 0 , \qquad\qquad\quad x \notin (0,c) , \end{aligned} \tag{7.10}$$

we can make an even stronger statement. In fact, we know that if the common density of the X_i's is of the form (7.10), we have

$$X_{m+2:2m+3} \leq_L X_{m+1:2m+1} \tag{7.11}$$

for every $m = 0,1,2,\ldots$ (this was Example 6.7). Since $E(X_{m+1:2m+1}) = c/2$, $\forall\, m$, we can conclude that

$$\mathrm{var}(X_{m+1:2m+1}) \downarrow \text{ as } m \uparrow \tag{7.12}$$

(since $u(x) = x^2$ is convex).

If we turn to consider sample means rather than medians, then, assuming second moments exist, we have for each n,

$$\mathrm{var}\,\bar{X}_{n+1} \leq \mathrm{var}\,\bar{X}_n \tag{7.13}$$

(where $\bar{X}_n = \frac{1}{n} \sum_{i=1}^{n} X_i$). In fact $\bar{X}_n$ and $\bar{X}_{n+1}$ are Lorenz ordered.

To see this, we argue as follows. By exchangeability of the X_i's we have $E(X_i | X_1 + \ldots + X_{n+1}) = (X_1 + \ldots + X_{n+1})/(n+1)$ for $i=1,2,\ldots,n+1$. Thus,

$$\begin{aligned} E(\bar{X}_n | \bar{X}_{n+1}) &= \frac{1}{n} E(\sum_{i=1}^{n} X_i | \bar{X}_{n+1}) \\ &= \frac{1}{n} \sum_{i=1}^{n} E(X_i | X_1 + \ldots + X_{n+1}) \\ &= \bar{X}_{n+1} \ . \end{aligned} \tag{7.14}$$

Since $E(\bar{X}_{n+1}) = E(\bar{X}_n)$, we can apply Theorem 3.4 and conclude from (7.14) that $\bar{X}_{n+1} \leq_L \bar{X}_n$.

Thus, if we were considering use of $\bar{X}_n$ or $\bar{X}_{n+1}$ as estimates of $E(X_1) = \mu$ and had a loss function which was a convex function of the error of estimate, i.e., of the form $g(T - \mu)$ where T is the estimate and g is convex, then no matter what choice of convex g is deemed appropriate we would prefer $\bar{X}_{n+1}$ to $\bar{X}_n$.

Under what circumstances is the sample median less unequal in the Lorenz order sense than the sample mean? An answer to this question would be of interest in that it would identify the forms of the common distribution of the X_i's for which the median would be generally preferred to the mean for estimation of the center of the distribution. For example, in the case in which the common distribution of the X_i's is uniform $(0,\sigma)$, one may verify by a density crossing argument (Theorem 6.5) that

$$\bar{X}_3 \leq_L X_{2:3} \ . \tag{7.15}$$

Equation (7.15) does not always hold. For example, if the common distribution of the X_i's is discrete with

$$P(X = 0) = P(X = 2) = 1/7, \quad P(X = 1) = 5/7 \ , \tag{7.16}$$

then $\bar{X}_3$ and $X_{2:3}$ are not Lorenz comparable. Note that (7.16) defines a symmetric distribution, so symmetry alone is not enough to guarantee (7.15).

Intuitively, (7.15) should hold for light tailed distributions. See Arnold and Villaseñor (1986) for further discussion.

7.4 Reliability

Consider a complex system involving n independent components. Each component either functions or not. Denote by p_i the reliability of component i, i.e., the probability that the i'th component functions, i=1,2,...n. The probability that the system functions will be some function of $\underline{p}$, say $h(\underline{p})$, called the reliability of the system. For example, a series system which functions only if all components function will have as its reliability function $h(\underline{p}) = \prod_{i=1}^{n} p_i$. A k out of n system functions if any k of its n components are functioning. For such a system we have

$$h(\underline{p}) = P\left(\max_{\pi} \sum_{j=1}^{k} X_{\pi(j)} = k\right) \tag{7.17}$$

where the summation is over all permutations of length k of the integers 1,2,...,n and X_i is an indicator random variable of the event that component i is functioning (i=1,2,...,n). The expression on the right of (7.17) clearly is a function of $\underline{p}$ only, albeit a complicated function.

Suppose that we have limited resources and must use them to construct components. For a given system, should we try to construct all components to have equal reliabilities, or are there key components on which we should concentrate our resources? If $h(\underline{p})$ is a symmetric function of $\underline{p}$ (as in the series, parallel and k out of n systems), intuition suggests that equal values of the p_i's would be reasonable. A majorization result may be sought. Roughly, we might expect that increased inequality among the component reliabilities, the p_i's, will decrease overall system reliability. Or, maybe, it is the other way round; it depends on your intuition. Can we prove such a result? It turns out that a majorization result can be proved. But the appropriate parameterization does not involve p_i's, rather we should let

$$\lambda_i = - \log p_i \qquad i=1,2,\ldots n \; . \tag{7.18}$$

Pledger and Proschan (1971) show that, indeed, in any k out of n system ($k = 1,2,\ldots,n$), the system reliability is a Schur convex function of $\underline{\lambda}$.

To see that this is true, we argue as follows. Consider two vectors $(\lambda_1,\lambda_2,\ldots,\lambda_n)$ and $(\lambda_1',\lambda_2',\ldots,\lambda_n)$ which differ only in their first two coordinates, and without loss of generality assume $\lambda_1 < \lambda_2$, $\lambda_1' = \lambda_1 + \varepsilon$, $\lambda_2' = \lambda_2 - \varepsilon$ where ε is small. We must show $h(\underline{\lambda}) \geq h(\underline{\lambda}')$. Let δ_k denote the probability that at least k of components 3 through n are functioning, and let δ_{k-1}, δ_{k-2} denote the probability that exactly k-1 (respectively k-2) of components 3 though n are functioning. Conditioning on the number of components functioning among components 3 through n, we find

$$\begin{aligned} h(\underline{\lambda}) - h(\underline{\lambda}') = {} & \delta_k(1 - 1) \\ & + \delta_{k-1}[(p_1 + p_2 - p_1p_2) - (p_1' + p_2' - p_1'p_2')] \\ & + \delta_{k-2}(p_1p_2 - p_1'p_2') \end{aligned}$$

where $\lambda_i = - \log p_i$ and $\lambda_i' = - \log p_i'$. Since $\lambda_1 + \lambda_2 = \lambda_1' + \lambda_2'$, we have $p_1p_2 = p_1'p_2'$. So we can write

$$\begin{aligned} h(\underline{\lambda}) - h(\underline{\lambda}') &= \delta_{k-1}[p_1 + p_2 - p_1' - p_2'] \\ &= \delta_{k-1}[e^{-\lambda_1} + e^{-\lambda_2} - e^{-\lambda_1+\varepsilon} - e^{-\lambda_2-\varepsilon}] \end{aligned}$$

which is positive, since e^{-x} is a decreasing convex function.

Note that in order to compare $\underline{\lambda}$ and $\underline{\lambda}'$ in the sense of majorization, we must have $\sum_{i=1}^n \lambda_i = \sum_{i=1}^n \lambda_i'$ or equivalently $\prod_{i=1}^n p_i = \prod_{i=1}^n p_i'$ (where $p_i' = e^{-\lambda_i'}$). Thus, the Pledger-Proschan result tells us about orderings of system reliabilities of k out of n systems as functions of individual reliabilities subject to the constraint that the reliability of a series system constructed with the components is some fixed quantity.

7.5. Genetic selection

Suppose that we are in the cattle breeding business, and we wish to develop a strain of cattle with a high beef yield. A reasonable approach to this problem involves culling in each generation and saving for further breeding purposes only the most meaty cattle. It is frequently deemed appropriate to model such a situation with a random effects linear model. Thus, in a particular generation we will have, say, m families of cattle of sizes $k_1, k_3, \ldots, k_m$. The meat yield of the i'th member of the j'th family is represented by

$$X_{ij} = \mu + A_i + E_{ij}, \quad \begin{matrix} i=1,2,\ldots,m \\ j=1,2,\ldots,k_i \end{matrix}, \tag{7.19}$$

where the A_i's are assumed to be i.i.d. normal $(0,\sigma_A^2)$ random variables, and the E_{ij} are i.i.d. normal $(0,\sigma^2)$ random variables. This is a classical intraclass correlation model. The culling scheme involves retaining the m' animals corresponding to the m' largest observed values of the X_{ij}'s. See Rawlings (1976), Hill (1976, 1977) and Tong (1982) for a more precise and detailed description of the phenomena in question. Rawlings and Hill assumed $k_i = k$, $i=1,1,\ldots,m$. We are here focussing on Tong's results which are concerned with the effect of heterogeneity of the k_i's. Tong restricts attention to the largest X_{ij} which we here denote by Z. He first observes that for every z, the quantity $P(Z \leq z)$ is a monotone increasing function of σ_A^2/σ^2. This is intuitively plausible. If we think of the extreme case when σ_A^2 is very large, then we are really only effectively dealing with the maximum of m rather than $\sum_{i=1}^{m} k_i = n$ variables, and Z will tend to be smaller.

The other result obtained by Tong is that increasing the "variability" of the vector of family sizes also tends to make Z stochastically smaller. In this setting we measure "variablility" in the sense of majorization. Specifically, consider two data sets of the form (7.19) denotes by $\{X_{ij}\}$ and $\{X'_{ij}\}$. Let the corresponding vectors of family sizes be $\underline{k} = (k_1,\ldots,k_m)$ and $\underline{k}' = (k_1',\ldots,k_m')$ respectively. Assume that the total number of observations is the same in both data sets, i.e., $\sum_{i=1}^{m} k_i = \sum_{i=1}^{m} k_i' = n$ say. Denote the corresponding maxima by Z and Z'. Tong's result may then be stated in the form:

<u>Lemma 7.4</u>: If $\underline{k} \leq_M \underline{k}'$, then for every z, $P(Z \leq z) \leq P(Z' \leq z)$.

<u>Proof</u>: Referring back to the model (7.19), we may compute for any z,

$$P(X_{ij} \leq z | A_i = a) = \phi_z(a) \text{ say.}$$

Also note that, given $A_1,A_2,\ldots,A_m$, the X_{ij}'s are conditionally independent. It follows that

$$\begin{aligned} P(Z \leq z) &= P(X_{ij} \leq z \quad \forall\, i,j) \\ &= E(P(X_{ij} \leq z \quad \forall\, i,j | A_1,A_2,\ldots,A_m)) \\ &= E\{\prod_{i=1}^{m} [\phi_z(A_i)]^{k_i}\} \\ &= E\{\sum_{\pi} \prod_{i=1}^{m} [\phi_z(A_{\pi(i)})]^{k_i}\}/m! \end{aligned} \tag{7.20}$$

where the summation is over all permutations of the integers $1,2,\ldots,m$. Analogously,

$$P(Z' \leq z) = E\{\sum_{\pi} \prod_{i=1}^{m} [\phi_z(A_{\pi(i)})]^{k_i'}\}/m! \; . \tag{7.21}$$

If $\underline{k} \leq_M \underline{k}'$, then the expressions inside the expectations in (7.20) and (7.21) are ordered by Muirhead's theorem (Theorem 2.11). The result then follows.

In the above Lemma it is evidently sufficient to have the joint distribution of $(A_1,A_2,\ldots,A_m)$ be exchangeable, the A_i's do not have to be independent. It is also evident that normality plays no role in the result.

An interesting open question is whether this majorization result can be extended to cover the case of the largest k of the X_{ij}'s rather than just the largest one. Other possible extensions are mentioned by Tong.

7.6. Large interactions

Bechhofer et al (1977) encountered a majorization relationship in the study of $2 \times c$ two way analysis of variance. They use the following constrained maximization result provided by Kemperman (1973).

Lemma 7.5: Let $\underline{x} \in \mathbb{R}^n$ satisfy $\ell \le x_i \le u$, $i=1,2,\ldots,n$. Then

$$\underline{x} \le_M (\ell,\ell,\ell,\ldots,\ell,\tau,u,u,\ldots,u) = \underline{y}$$

where there are k ℓ's and $n-k-1$ u's and k and $\tau \in [\ell,u]$ are uniquely determined by the requirement that $\sum_{i=1}^{n} x_i = \sum_{i=1}^{n} y_i$.

Proof: Since $\ell \le x_i \le u$, it follows that, for $j \le k$, $\sum_{i=1}^{j} x_{i:n} \ge \sum_{i=1}^{j} y_{i:n}$ and for $j > k$, that $\sum_{i=j+1}^{n} x_{i:n} \le \sum_{i=j+1}^{n} y_{i:n}$. Thus, for all $j \le n$, $\sum_{i=1}^{j} x_{i:n} \ge \sum_{i=1}^{j} y_{i:n}$, i.e., $\underline{x} \le_M \underline{y}$.

Now consider a two way fixed effects analysis of variance situation with 2 rows and J columns with K observations per cell. Thus, our data is of the form

$$X_{ijk} = \mu + \alpha_i + \beta_j + \gamma_{ij} + E_{ijk}, \quad \begin{array}{l} i=1,2 \\ j=1,2,\ldots,J \\ k=1,2,\ldots,K \end{array}$$

where the E_{ijk}'s are i.i.d. $N(0,\sigma^2)$ where σ^2 is assumed known. As usual, we assume our parameters satisfy certain constraints:

$$0 = \sum_{i=1}^{2} \alpha_i = \sum_{j=1}^{J} \beta_j = \sum_{i=1}^{2} \gamma_{ij} = \sum_{j=1}^{J} \gamma_{ij} . \tag{7.22}$$

Suppose that we are interested in identifying the cell whose interaction parameter γ_{ij} is largest. We wish to derive a selection procedure which will identify the largest interaction cell with high probability when there is some discernible interaction and when one cell interaction is discernibly larger than the others. Specifically, we want a rule that will select the cell corresponding to the largest interaction parameter which without loss of generality will be the (1,1) cell with probability at least P^* when $\gamma_{11} \geq \Delta^*$ and $\gamma_{ij} < \Delta^*-\delta^*$, $(i,j) \neq (1,1)$ where P^*, Δ^* and δ^* are specified in advance by the experimenter. This is a typical formulation of a ranking and selection problem (see e.g. Bechhofer, Kiefer and Sobel, 1968).

A plausible decision rule is one which selects as the cell with largest interaction that cell whose observed interaction, namely,

$$Z_{ij} = X_{ij.} - X_{i..} - X_{.j.} + X_{...} \tag{7.23}$$

is largest. As usual, in (7.23) the dots indicate averaging over missing subscripts. For a configuration of interaction parameters $\underline{\gamma}$ satisfying

$$\gamma_{11} \geq \Delta^*$$

and (7.24)

$$\gamma_{ij} \leq \Delta^* - \delta^*, \ (i,j) \neq (1,1) ,$$

the probability of correct selection using the above described rule is simply

$$P_{\underline{\gamma}}(Z_{1,1} > Z_{i,j} \quad \forall\, (i,j) \neq (1,1)) . \tag{7.25}$$

We want to determine the sample size necessary to assure us that the probability of correct selection given by (7.25) is at least equal to P^*. It is evident that we can easily achieve this goal if K is enormous. What is the smallest value of K which will suffice? Let Γ denote the set of all interaction arrays $\underline{\gamma}$ which satisfy (7.24). Is there a least favorable $\underline{\gamma}$ in Γ (i.e. one for which the probability of correct selection (7.25) is smallest)? There is, and it can be identified by a majorization argument. Observe that an array $\underline{\gamma}$ will belong to Γ provided that

$$\gamma_{11} \geq \Delta^*$$

and

$$|\gamma_{ij}| \leq \gamma_{11} - \delta^*, \quad j=2,\ldots,J . \tag{7.26}$$

Also note that $\sum_{j=2}^{J} \gamma_{1j} = -\gamma_{11}$. Expression (7.26) describes a coordinatewise bounded collection of vectors of dimension $(J - 1)$ to which

Kemperman's result (Lemma 7.5) can be applied. The vector $\tilde{\underline{\gamma}} = (\gamma_{12},\gamma_{13},\ldots,\gamma_{1J})$ is majorized by the vector

$$\tilde{\underline{\gamma}}^* = (\ell,\ell,\ldots,\ell,\tau,u,u,\ldots u)$$

described in Lemma 7.5, with

$$\ell = -(\gamma_{11} - \delta^*) ,$$

$$u = (\gamma_{11} - \delta^*)$$

and k and τ determined such that $\sum_{j=2}^{J} \gamma_{1j} = k\ell + \tau + (n - k - 1)u$. The vector $\tilde{\underline{\gamma}}^*$ can be used in conjunction with the constraints (7.22) to determine an array $\underline{\gamma}^*$ in Γ (with $\gamma_{11}^* = \gamma_{11}$). A non-trivial application of Theorem 2.8 yields the result that the probability of correct selection for a fixed value of γ_{11} is a Schur concave function of $\tilde{\underline{\gamma}} = (\gamma_{12},\ldots,\gamma_{1J})$. It follows then that

$$P_{\underline{\gamma}}(Z_{11} > Z_{ij} \;\forall\, (i,j) \neq (1,1))$$
$$\geq P_{\underline{\gamma}^*}(Z_{11} > Z_{ij} \;\forall\, (i,j) \neq (1,1)) .$$

It remains only to verify that the probability of correct selection is a monotone increasing function of γ_{11}, and then we conclude that the least favorable configuration is of the form $\underline{\gamma}^*$ with $\gamma_{11}^* = \Delta^*$. In principle, we can then determine by numerical integration the minimal value of K necessary to achieve the desired level P* for the probability of correct selection for any $\underline{\gamma}$ in Γ.

7.7. Unbiased tests

Suppose that we have a data set $\underline{X}$ whose distribution depends on a k-dimensional parameter $(\theta_1,\theta_2,\ldots,\theta_k)$. Not infrequently we are interested in testing the hypothesis of homogeneity, i.e., H: $\theta_1 = \theta_2 = \ldots = \theta_k$. The test will often be of the form: reject H if $\phi(\underline{X}) \geq c$. Consequently, the power function will be

$$\beta(\underline{\theta}) = P_{\underline{\theta}}(\phi(\underline{X}) \geq c).$$

It is often the case that ϕ is a Schur convex function of $\underline{x}$, and consequently, that $\beta(\underline{\theta})$ is a Schur convex function of $\underline{\theta}$. This information may permit us to conclude that our test is unbiased. A specific case in which this program works perfectly is that in which X has a multinomial $(N,\underline{\theta})$ distribution and H: $\theta_1 = \theta_2 = \ldots = \theta_k = 1/k$. In this setting the likelihood ratio test is of the form: reject H if $\sum_{i=1}^{k} X_i^2 \geq c$. This is clearly a Schur convex function of $\underline{X}$, and the test is verified to be unbiased (cf. Perlman and Rinott, 1977). The program is also feasible to verify unbiasedness of the likelihood ratio test of sphericity of a multivariate normal population and certain invariant tests of equality of mean vectors in multivariate analysis of variance.

7.8. Summation modulo m

Suppose X and Y are independent random variables with possible values 0,1,2,...,m-1. Define $Z = X \oplus Y$ where the sybol $\oplus$ denotes addition modulo m. Define vectors $\underline{p}$, $\underline{q}$, and $\underline{r}$ by

$$p_i = P(X = i)\ ,$$

$$q_i = P(Y = i)\ ,$$

$$r_i = P(Z = i)\ ,$$

where i=0,1,2,...,m-1.

Lemma 7.6: $\underline{r} \leq_M \underline{p}$ and $\underline{r} \leq_M \underline{q}$.

Proof: By conditioning we can verify that

$$\underline{r} = P\underline{q}$$

where P is the circulant matrix

$$\begin{bmatrix} p_0 & p_{m-1} & p_{m-2} & \cdots & p_2 & p_1 \\ p_1 & p_0 & p_{m-1} & \cdots & p_3 & p_2 \\ \cdot & & & & & \\ \cdot & & & & & \\ \cdot & & & & & \\ p_{m-1} & p_{m-2} & \cdots & \cdots & p_1 & p_0 \end{bmatrix}$$

Evidently, P is doubly stochastic and, thus, by the HLP Theorem 2.1, we conclude $\underline{r} \leq_M \underline{q}$. Analogously, $\underline{r} \leq_M \underline{p}$.

Lemma 7.6 can be generalized to cover random variables assuming values in an arbitrary finite group (rather than just the group of non-negative integers under summation modulo m). Marshall and Olkin (1979, p. 383-4) give the general expression. Brown and Solomon (1976) focussed on the case in which the group consisted of vectors of non-negative integers under coordinatewise summation modulo m.

The problem is of interest in the context of pseudo-random number generators. If our pseudo random number generator generates integers 0,1,2,...,9 with non-uniform probabilities, our Lemma tells us we may improve things (i.e. obtain more uniformly distributed random digits) by summing successively generated digits modulo 10. Or we might combine outputs from distinct random number generators. If this is done by addition modulo 10, our Lemma guarantees that the output will be at least as uniformly distributed and, most likely, more uniformly distributed than was the output of either individual generator.

The improvement can be striking. For example if

$$\underline{p} = (.07, .19, .02, .11, .06, .13, .01, .18, .09, .14)$$

and

$$\underline{q} = (.13, .07, .02, .11, .17, .13, .07, .02, .11, .17),$$

then from (7.27) we find

$$\underline{r} = (.1, .1, .1, .1, .1, .1, .1, .1, .1, .1) .$$

Examples such as this are discussed in Arnold (1979) in a related context.

7.9. Forecasting

Every night on TV Channel 5, Bert announces the probability of rain for the next day. Every night on Channel 11, Wilbur gives out his probability of rain. Is Bert better than Wilbur? How should we appropriately determine if one is the better forecaster? De Groot and Fienberg have studied this phenomenon in a series of papers. A representative reference addressed to the general reader is De Groot and Fienberg (1983). The concept of majorization plays a central role. In the present section the scenario will only be sketched along the lines suggested by De Groot and Fienberg.

A particular forecaster is asked each day to give us his subjective probability say x that there will be a measurable amount of precipitation at a certain location. At the end of each day we can observe whether or not it did rain. Denote by X the random prediction of our forecaster (most easily visualized in a long run frequency sense). Denote by Y an indicator random variable. Y assumes the value 1 if it rains, 0 otherwise. The performance of our forecaster is summarized by the density function $\nu(x)$ of the random variable X and for each possible value x of X, the conditional probability of rain given that the forecaster's prediction is x, denoted by $\rho(x)$. Thus

$$P(X \varepsilon A) = \int_A \nu(x)\, dx \tag{7.28}$$

and

$$P(Y = 1 | X = x) = \rho(x) \ . \tag{7.29}$$

The forecaster is said to be well-calibrated if $\rho(x) \equiv x$ (such a forecaster is sometimes said to be perfectly reliable). If he says there is a 50% chance of rain, then indeed there is a 50% chance of rain. Any coherent forecaster who updates his subjective probability based on experience assuming Bayes theorem, will be asymptotically well calibrated (see e.g. Dawid (1982)). It is usually deemed appropriate to restrict attention to well calibrated forecasters. Clearly however there are differences between well calibrated forecasters. If in a given town rain occurs on about 15% of the days during the year, then a forecaster who every day announces x = .15 will be well calibrated but somewhat boring and of questionable utility. At the other extreme a forecaster who always announces either x = 0 or x = 1 and who is always right is also well-calibrated (and displays amazing skill).

Comparisons between well calibrated forecasters are made in terms of a partial ordering known as refinement. Consider two well calibrated forecasters A and B whose forecasts X_A and X_B have corresponding density functions ν_A and ν_B (as in (7.28)). Since both are well calibrated we have (cf. (7.29))

$$\rho_A(x) = \rho_B(x) = x, \quad \forall\ x \in [0,1].$$

A stochastic transformation is a function $h(y|x)$ defined on $[0,1] \times [0,1]$ satisfying

$$\int_0^1 h(y|x)\, dy = 1, \quad \forall\ x \in [0,1]\ . \tag{7.30}$$

Forecaster A is said to be at least as refined as forecaster B if there exists a stochastic transformation $h(y|x)$ for which

$$\nu_B(y) = \int_0^1 h(y|x)\ \nu_A(x)\ dx, \ \forall\ y \in [0,1] \tag{7.31}$$

and

$$y\nu_B(y) = \int_0^1 h(y|x)\ x\ \nu_A(x)\ dx, \ \forall\ y \in [0,1]\ . \tag{7.32}$$

We denote by μ, the long run frequency of rainy days. For any well calibrated forecaster we must have

$$\int_0^1 x\ \nu(x)\ dx = \mu.$$

Any density ν with mean μ can be identified with a well calibrated forecaster. Thus our refinement partial order may be thought of as being defined on the class of all random variables with range [0,1] and mean μ. In this context we can allow the prediction random variables X to be discrete with obvious modifications in our earlier discussion which assumed absolutely continuous prediction variables (just replace integral signs by summation signs).

Inspection of conditions (7.31) and (7.32), in either the discrete or absolutely continuous case, yields the interpretation that forecaster A (with prediction variable X_A) is at least as refined as B (with prediction variable X_B) if and only if there exist random variables X_A' and X_B' defined on some convenient probability space with

$$\begin{aligned} X_A' &\overset{d}{=} X_A\ , \\ X_B' &\overset{d}{=} X_B \end{aligned} \tag{7.33}$$

and

$$E(X'_A \mid X'_B) = X'_B \ . \tag{7.34}$$

Thus the refinement ordering, on random variables with support [0,1] and mean μ, is exactly identifiable with Lorenz ordering on that restricted class of random variables (recall Theorem 3.4). Since our random variables have identical means it can also be identified, if desired, with the second order stochastic dominance partial order (as defined in Chapter 6).

If (7.33) and (7.34) hold then we know that

$$E(u(X_A)) \geq E(u(X_B)) \tag{7.35}$$

for any convex function u. In the forecasting setting this has an interpretation in terms of what are known as strictly proper scoring rules.

A scoring rule is a device for comparing forecasters (even it they are not well calibrated). Suppose that the forecaster's prediction is x. If rain does occur he receives a score of $g_1(x)$, if it does not rain he receives a score of $g_2(x)$. The expected score is, using (7.28) and (7.29), thus

$$S = E[\rho(x)g_1(x) + [1 - \rho(x)]g_2(x)] \ . \tag{7.36}$$

It is reasonable to assume that g_1 is an increasing function and g_2 is a decreasing function. A scoring rule is said to be strictly proper if for each value of y in the unit interval the function

$$y\, g_1(x) + (1 - y)\, g_2(x) \tag{7.37}$$

is maximized only when x = y. Such a rule encourages the predictor to announce his true subjective probability of rain. It is not difficult to verify that if g_1 is increasing, g_2 is decreasing and the scoring rule is strictly proper then the function

$$x\, g_1(x) + (1 - x)\, g_2(x) \tag{7.38}$$

is strictly convex. Referring to (7.36), if we consider two well calibrated forecasters A and B (for whom $\rho_A(x) = \rho_B(x) = x$) where A is at least as refined as B, then using any proper scoring rule forecaster A will receive a higher score than forecaster B (unless $X_A \overset{d}{=} X_B$ in which case they clearly receive identical scores).

Recently De Groot and Fienberg (1986) have extended these concepts to allow for comparison of "multivariate" forecasters. Thus instead of predicting (rain)-(no-rain) the forecaster might predict a finite set of temperature ranges. The prediction random variable X becomes s dimensional (assuming values in the simplex $x_i \geq 0$, $\sum_{i=1}^{s} x_i = 1$). The concepts of refinement, well calibration and scoring rules continue to be meaningful in this more complex setting.

7.10. Ecological diversity

Consider a closed ecological community (an island in the Pacific, for example). Following an exhaustive study, representatives of s different species of insects are found on the island. For $i=1,2,\ldots,s$ let n_i denote the abundance of species i, the number of insects of that species on the island. For $i=1,2,\ldots,s$ define the relative abundance π_i by

$$\pi_i = n_i/(\sum_{j=1}^{s} n_j) . \tag{7.39}$$

The community can be identified with its relative abundance profile $(\pi_1,\ldots,\pi_s)$.

Evidently some communities are more diverse than others. What is an acceptable ordering or partial ordering of relative abundance profiles which will capture the spirit of the concept of diversity? Several diversity measures have been proposed in the literature. Patil and Taillie (1982), taking on the ecological role played in the economics context by Dalton, sought basic principles of diversity which might lead to a widely accepted diversity order. They were led in this fashion to what they call the intrinsic diversity ordering.

Suppose that $\underline{\pi}^{(1)}$ and $\underline{\pi}^{(2)}$ are species abundance profiles of two communities involving (without loss of generality) the same s species in both communities. When is it appropriate to say that $\underline{\pi}^{(1)}$ is more diverse then $\underline{\pi}^{(2)}$?

The diversity of an abundance profile $\underline{\pi}$ is deemed to be a function of the ordered components of $\underline{\pi}$; viz. $\pi_{1:s}, \pi_{2:s}, \ldots, \pi_{s:s}$. Thus a population with 25% butterflies, 60% ants and 15% beetles is deemed to be equally diverse as one containing 60% butterflies, 15% ants and 25% beetles. The intrinsic diversity profile (IDP) is actually defined in terms of the decreasing order statistics of π, i.e. $\pi_{(1:s)} \geq \pi_{(2:s)} \geq \cdots \geq \pi_{s:s}$. It is a plot of the points $\{(\sum_{j=1}^{i} \pi_{(j:s)}, i/s)\}_{i=1}^{s}$. The curve is completed by linear interpolation. If all the species are equally represented the IDP will be a 45° line. This is taken to be the case of maximum diversity. If one species is extremely numerous [i.e. $\pi_{(1:s)} \approx 1$], the curve will be very low and diversity is judged to be small.

Community $\underline{\pi}^{(1)}$ will be judged to be intrinsically more diverse than community $\underline{\pi}^{(2)}$ if the IDP of $\underline{\pi}^{(1)}$ is uniformly above the IDP of $\underline{\pi}^{(2)}$ and we may write

$$\underline{\pi}^{(1)} \underset{\text{IMD}}{>} \underline{\pi}^{(2)} .$$

It is interesting to view how this intrinsic diversity ordering ranks uniform distributions over different numbers of species. Suppose $\underline{\pi}_1$ has K_1 non-zero components all equal to $1/K_1$ while $\underline{\pi}_2$ has K_2 non-zero components, all equal to $1/K_2$. If $K_1 > K_2$, then $\underline{\pi}_1$ surely exhibits more diversity than $\underline{\pi}_2$. Inspection of the corresponding IDP's confirms that in this case $\underline{\pi}_1 \underset{\text{IMD}}{>} \underline{\pi}_2$.

Referring to the definition of an intrinsic diversity profile and the definition of majorization (in $\mathbb{R}^s$) we see that they are intimately related. In fact

$$\underline{\pi}^{(1)} \underset{\text{IMD}}{\geq} \underline{\pi}^{(2)} <=> \underline{\pi}^{(1)} \leq_M \underline{\pi}^{(2)}. \qquad (7.40)$$

In the light of (7.40), summary measures of diversity should be Schur concave functions of $\underline{\pi}$ in order to respect the intrinsic diversity ordering. Typical examples of such measures are:

$$\text{(Shannon's measure) } D_1(\underline{\pi}) = - \sum_{i=1}^{s} \pi_i \log \pi_i \ , \qquad (7.41)$$

$$\text{(Simpson's measure) } D_2(\pi) = 1 - \sum_{i=1}^{s} \pi_i^2 \qquad (7.42)$$

Many diversity measures are interpretable as coefficients of expected rarity. Imagine that each species has a measure of rarity attached to it. The simplest case occurs when there exists a rarity function $R(\pi)$ defined on $[0,1]$ such that the rarity of an individual belonging to a species with relative abundance π in the community is equal to $R(\pi)$. Imagine we randomly select an individual from the community. The expected rarity of the individual so chosen is then given by

$$(ER)(\underline{\pi}) = \sum_{i=1}^{s} \pi_i R(\pi_i) \ . \qquad (7.43)$$

Such a measure will be Schur concave, and hence will respect the intrinsic diversity order, if R is concave. Both the Shannon (7.41) and the Simpson (7.42) indices are of the form (7.43) for suitable concave functions $R(\pi)$.

Patil and Taillie (1982) suggest several alternative diversity profiles. They also present interpretations of many diversity indices in terms of random encounters (inter-species and intra-species).

REFERENCES

Aggarwal, V. and Singh, R. (1984) On optimum stratification with proportional allocation for a class of Pareto distributions. Communications in Statistics A13, 3107-3116.

Arnold, B. C. (1979) Non-uniform decompositions of uniform random variables under summation modulo m. Bollettino Unione Matematica Italiana (5), 16-A, 100-102.

Arnold, B. C. (1983) Pareto Distributions. International Cooperative Publishing House. Fairland, Maryland.

Arnold, B. C. (1986) Inequality attenuating and inequality preserving weightings. Technical Report #146, University of California, Riverside.

Arnold, B. C., Brockett, P. L., Robertson, C. A. and Shu, B.-Y. (1987) Generating ordered families of Lorenz curves by strongly unimodal distributions. Journal of Business and Economic Statistics, to appear.

Arnold, B. C. and Villaseñor, J. A. (1984) Some examples of fitted general quadratic Lorenz curves. Technical Report #130, University of California, Riverside.

Arnold, B. C. and Villaseñor, J. A. (1985) Inequality preserving and inequality attenuating transformations. Technical Report, Colegio de Postgraduados, Chapingo, Mexico.

Arnold, B. C. and Villaseñor, J. A. (1986) Lorenz ordering of means and medians. Statistics and Probability Letters 4, 47-49.

Bechhofer, R. E., Kiefer, J. and Sobel, M. (1968) Sequential identification and ranking procedures. University of Chicago Press, Chicago, Illinois.

Bechhofer, R. E., Santer, T. J. and Turnbull, B. W. (1977) Selecting the largest interaction in two-factor experiment. In Statistical Decision Theory and Related Topics II (S. S. Gupta and D. S. Moore, eds.), pp. 1-18. Academic Press, New York.

Birkhoff, G. (1946) Tres observaciones sobre el algebra lineal. Univ. Nac. Tucuman Rev. Ser. A, 5, 147-151.

Brown, M. and Solomon, H. (1976) On a method of combining pseudo-random number generators. Technical Report No. 233, Department of Statistics, Stanford University, Stanford, California.

Dalton, H. (1920) The measurement of the inequality of incomes. Economic Journal 30, 348-361.

Dawid, A. P. (1982) The well-calibrated Bayesian. J. American Statistical Association 77, 605-610.

De Groot, M. H. and Fienberg, S. E. (1983) The comparison and evaluation of forecasters. The Statistician 32, 12-22.

De Groot, M. H. and Fienberg, S. E. (1986) Comparing probability forecasters: Basic binary concepts and multivariate extensions, in P. K. Goel and A. Zellner, eds., Bayesian inference and decision techniques, Essays in honor of B. de Finetti, North Holland, Amsterdam, 247-264.

Farahat, H. K. and Mirsky, L. (1960) Permutation endomorphisms and refinement of a theorem of Birkhoff. Proc. Cambridge Philos. Soc. 56, 322-328.

Feller, W. (1971) An Introduction to Probability Theory and Its Applications, 2nd ed., Vol. 2. John Wiley and Sons, Inc., New York.

Fellman, J. (1976) The effect of transformations on Lorenz curves. Econometrica 44, 823-824.

Fishburn, P. C. (1976) Continua of stochastic dominance relations for bounded probability distributions. Journal of Mathematical Economics 3, 295-311.

Fishburn, P. C. (1980) Continua of stochastic dominance relations for unbounded probability distributions. Journal of Mathematical Economics 7, 271-285.

Gastwirth, J. L. (1971) A general definition of the Lorenz curve. Econometrica 39, 1037-1039.

Hardy, G. H., Littlewood, J. E. and Polya, G. (1929) Some simple inequalities satisfied by convex functions. Messenger Math. 58, 145-152.

Hardy, G. H., Littlewood, J. E. and Polya, G. (1959) Inequalities. Cambridge University Press, London and New York.

Hill, W. G. (1976) Order statistics of correlated variables and implications in genetic selection programmes. Biometrics 32, 889-902.

Hill, W. G. (1977) Order statistics of correlated variables and implications in genetic selection programmes II, Response to selection. Biometrics 33, 703-712.

Hollander, M., Proschan, F. and Sethuraman, J. (1977) Functions decreasing in transposition and their applications in ranking problems. Annals of Statistics 5, 722-733.

Horn, A. (1954) Doubly stochastic matrices and the diagonal of a rotation matrix. American Journal of Mathematics 76, 620-630.

Kakwani, N. C. (1980) Income Inequality and Poverty, Methods of Estimation and Policy Applications. Oxford University Press, New York.

Karamata, J. (1932) Sur une inegalite relative aux fonctions convexes. Publ. Math. Univ. Belgrade 1, 145-148.

Kemperman, J.H.B. (1973) Moment problems for sampling without replacement, I, II, III. Nederl. Akad. Wetensch. Proc., Ser. A, 76, 149-164, 165-180, and 181-188.

Lorenz, M. O. (1905) Methods of measuring the concentration of wealth. Publication of the American Statistical Association 9, 209-219.

Mahfoud, M. and Patil, G. P. (1982) On weighted distributions. In Statistics and Probability, Essays in honor of C. R. Rao, G. Kallianpur et al. (eds.), North Holland, Amsterdam, 479-492.

Marshall, A. W. and Olkin, I. (1974) Majorization in multivariate distributions. Annals of Statistics 2, 1189-1200.

Marshall, A. W. and Olkin, I. (1979) Inequalities: Theory of Majorization and Its Applications. Academic Press, New York.

Marshall, A. W., Olkin, I. and Proschan, F. (1967) Monotonicity of ratios of means and other applications of majorization. In Inequalities (O. Shisha, ed.), pp. 177-190. Academic Press, New York.

Mirsky, L. (1958) Matrices with prescribed characteristic roots and diagonal elements. J. London Math. Soc. 33, 14-21.

Muirhead, R. F. (1903) Some methods applicable to identities and inequalities of symmetric algebraic functions of n letters. Proceedings of Edinburgh Mathematical Society 21, 144-157.

Nevius, S. E., Proschan, F. and Sethuraman, J. (1977) Schur functions in statistics, II. Stochastic majorization. Annals of Statistics 5, 263-273.

Nygard, F. and Sandstrom, A. (1981) Measuring Income Inequality. Almqvist and Wiksell International, Stockholm.

Patil, G. P. and Taillie, C. (1982) Diversity as a concept and its measurement. Journal of the American Statistical Association 77, 548-561.

Perlman, M. D. and Rinott, Y. (1977) On the unbiasedness of goodness-of-fit tests. Unpublished manuscript.

Pledger, G. and Proschan, F. (1971) Comparisons of order statistics and of spacings from heterogeneous distributions. In Optimizing Methods in Statistics (J. S. Rustagi, ed.), pp. 89-113. Academic Press, New York.

Rao, C. R. (1965) On discrete distributions arising out of methods of ascertainment. In Classical and Contagious Discrete Distributions, (G. P. Patil, ed.), pp. 320-332. Pergamon Press, Calcutta.

Rawlings, J. O. (1976) Order statistics for a special case of unequally correlated multinomial variates. Biometrics 32, 875-887.

Samuelson, P. A. (1965) A fallacy in the interpretation of the alleged constancy of income distribution. Rivista Internazionale di Scienze Economiche e Commerciali 12, 246-253.

Schur, I. (1923) Uber eine klasse von mittelbildungen mit anwendungen die determinaten. Theorie Sitzungsber Berlin Math. Gesellschaft 22, 9-20.

Shaked, M. (1982) Dispersive ordering of distributions. Journal of Applied Probability 19, 310-320.

Strassen, V. (1965) The existence of probability measures with given marginals. Annals of Mathematical Statistics 36, 423-439.

Taguchi, T. (1972a) On the two-dimensional concentration surface and extensions of concentration coefficient and Pareto distribution to the two dimensional case-I. Annals of Institute of Statistical Mathematics 24, 355-382.

Taguchi, T. (1972b) On the two-dimensional concentration surface and extensions of concentration coefficient and Pareto distribution to the two dimensional case-II. Annals of Institute of Statistical Mathematics 24, 599-619.

Taillie, C. (1981) Lorenz ordering within the generalized gamma family of income distributions. In Statistical Distributions in Scientific Work (C. Taillie et al. (eds.)), Vol. 6, Reidel, Dordrecht-Holland, 181-192.

Tong, Y. L. (1982) Some applications of inequalities for extreme order statistics to a genetic selection problem. Biometrics 38, 333-339.

Whitmore, G. A. and M. C. Findlay, eds., (1978) Stochastic Dominance: An Approach to Decision-Making Under Risk. D. C. Heath, Lexington, MA.

Wold, H. (1935) A study on the mean difference, concentration curves and concentration ratio. Metron 12, 39-58.

Yang, H.-J. (1982) On the variances of median and some other statistics. Bulletin of the Institute of Mathematics, Academia Sinica 10, 197-204.

AUTHOR INDEX

SUBJECT INDEX

Lecture Notes in Statistics

Vol. 26: Robust and Nonlinear Time Series Analysis. Proceedings, 1983. Edited by J. Franke, W. Härdle and D. Martin. IX, 286 pages. 1984.

Vol. 27: A. Janssen, H. Milbrodt, H. Strasser, Infinitely Divisible Statistical Experiments. VI, 163 pages. 1985.

Vol. 28: S. Amari, Differential-Geometrical Methods in Statistics. V, 290 pages. 1985.

Vol. 29: Statistics in Ornithology. Edited by B.J.T. Morgan and P.M. North. XXV, 418 pages. 1985.

Vol. 30: J. Grandell, Stochastic Models of Air Pollutant Concentration. V, 110 pages. 1985.

Vol. 31: J. Pfanzagl, Asymptotic Expansions for General Statistical Models. VII, 505 pages. 1985.

Vol. 32: Generalized Linear Models. Proceedings, 1985. Edited by R. Gilchrist, B. Francis and J. Whittaker. VI, 178 pages. 1985.

Vol. 33: M. Csörgő, S. Csörgő, L. Horváth, An Asymptotic Theory for Empirical Reliability and Concentration Processes. V, 171 pages. 1986.

Vol. 34: D.E. Critchlow, Metric Methods for Analyzing Partially Ranked Data. X, 216 pages. 1985.

Vol. 35: Linear Statistical Inference. Proceedings, 1984. Edited by T. Caliński and W. Klonecki. VI, 318 pages. 1985.

Vol. 36: B. Matérn, Spatial Variation. Second Edition. 151 pages. 1986.

Vol. 37: Advances in Order Restricted Statistical Inference. Proceedings, 1985. Edited by R. Dykstra, T. Robertson and F.T. Wright. VIII, 295 pages. 1986.

Vol. 38: Survey Research Designs: Towards a Better Understanding of Their Costs and Benefits. Edited by R.W. Pearson and R.F. Boruch. V, 129 pages. 1986.

Vol. 39: J.D. Malley, Optimal Unbiased Estimation of Variance Components. IX, 146 pages. 1986.

Vol. 40: H.R. Lerche, Boundary Crossing of Brownian Motion. V, 142 pages. 1986.

Vol. 41: F. Baccelli, P. Brémaud, Palm Probabilities and Stationary Queues. VII, 106 pages. 1987.

Vol. 42: S. Kullback, J.C. Keegel, J.H. Kullback, Topics in Statistical Information Theory. IX, 158 pages. 1987.

Vol. 43: B.C. Arnold, Majorization and the Lorenz Order: A Brief Introduction. VI, 122 pages. 1987.

Lecture Notes in Statistics